Springer Remote Sensing/Photogrammetry

More information about this series at http://www.springer.com/series/10182

Yosio Edemir Shimabukuro
Flávio Jorge Ponzoni

Spectral Mixture for Remote Sensing

Linear Model and Applications

Springer

Yosio Edemir Shimabukuro
Remote Sensing Division
National Institute for Space Research
São José dos Campos, São Paulo, Brazil

Flávio Jorge Ponzoni
Remote Sensing Division
National Institute for Space Research
São José dos Campos, São Paulo, Brazil

ISSN 2198-0721 ISSN 2198-073X (electronic)
Springer Remote Sensing/Photogrammetry
ISBN 978-3-030-02016-3 ISBN 978-3-030-02017-0 (eBook)
https://doi.org/10.1007/978-3-030-02017-0

Library of Congress Control Number: 2018958512

This Springer imprint is published by the registered company Springer Nature Switzerland AG
The registered company address is: Gewerbestrasse 11, 6330 Cham, Switzerland

Foreword

Before introducing the book *Spectral Mixture for Remote Sensing: Linear Model and Applications*, I feel impelled to speak about the authors and my relationship with them. Both authors are of the utmost personal and intellectual integrity, pioneers in the implementation of remote sensing in Brazil. Both have solid academic education from institutions of international renown. I had the privilege of being the Master degree advisor of the first author, Yosio Shimabukuro, and have hired the second author, Flávio Ponzoni, for the Remote Sensing Division of the National Institute for Space Research (INPE). I spent most of my professional life with both, as co-worker and scientific research collaborator. I followed their professional growth and had the satisfaction to see that they exceeded the one who could be considered their master one day. They published several books and book chapters with prestigious publishers, and they also published several scientific articles of international impact, which elevated the first author to the 1A Researcher level of the National Council for Scientific and Technological Development (CNPq), considered the highest recognition of a researcher by that institution.

The authors' experience includes the development of several models for application of remote sensing based on physical principles, aiming to the relevant applications for the knowledge and monitoring of Earth resources. This book describes in detail a model that allows transforming spectral mixture – into useful information for users. This book allows readers to navigate through the fundamentals of remote sensing since its origin. It describes the main satellites, sensors, and spectral indices used in the interpretation of images. It also presents the operational remote sensing projects, as well as the physical principles, cartographic bases, features, and image formats, to focus on linear spectral mixture model in detail.

This book presents the history of remote sensing since the use of aerial photographs to the modern concepts of information extraction exploring the various resolutions of the images, with emphasis on the concept of spectral mixture and how to decompose this mixture into its various fractions. It emphasizes, in a didactic way, the applications of this model in large projects, especially in the Amazon region where the dynamics of land use transformation is intense. Thus, the PRODES, DETER, PANAMAZONIA, and AMAZONICA projects are presented for the estimation of deforested and burned areas, based on several spatial and temporal resolutions of images, and presented at different scales.

By focusing on the main subject of the book, which is the decomposition of the elements of images in biophysical components familiar to every researcher as vegetation, soil, and water/shade, the authors present different mathematical approaches and justify the advantages of the linear model described in detail. The authors have a large experience in the implementation and application of this model, since the main author developed the linear spectral mixture model for the MSS and TM sensors of the Landsat satellite, in 1987, in his Ph.D. program at Colorado State University, USA. They describe the difficulties that users have working with "digital numbers" or "gray levels" of the images and the effect of the atmosphere in the geophysical or biophysical objects characterization.

After the conceptualization of the mixture model, the authors describe how it can be implemented in the major image processing systems such as SPRING, ENVI, and PCI. The description resembles an operating manual, with actual examples of fraction images of actual scenes. They emphasize the advantages of fraction images, such as the facility of interpretation, instead of using gray levels analysis. They also show how the model can be used to reduce the data dimensionality, for example, by transforming data from several sensor bands into three fractions (vegetation, soil, and water/shade).

This book progresses with a generic description of the main sensors/observation platforms of the Earth and concludes with a detailed description of how the products derived from the linear spectral mixture model are used in the operational project analysis steps from INPE, which include image segmentation of fraction images, classification, and matrix edition for presentation on various scales for dissemination of the results. The contribution of the fraction images was fundamental for the automation of these projects. They also proved their use in burned areas mapping and selective logging monitoring using the Landsat images in Mato Grosso state.

I am very confident that this book is a relevant contribution to the science and applications of remote sensing.

Getúlio Teixeira Batista
Ph. D. in Remote Sensing and Agriculture, Purdue University, 1981.
Editor of Ambiente & Água – an International Journal of Applied Science
Environmental Science Post-Graduate School, University of Taubaté
Taubaté, SP, Brazil

Preface

The development of remote sensing techniques was marked by three moments. The first of these was the improvement of photography that made possible the development of photogrammetry and photointerpretation techniques. The second came almost as an extension of these two forms of information extraction, now applied to images generated by electro-optical sensors placed onboard orbital platforms (satellites). Initially the applications explored approaches similar to those employed in the extraction of information from photographs and, subsequently, motivated by the recent discussions on climate change and the consequent needs to provide data on greenhouse gas emissions and other contributors to global warming. The third moment was characterized by more applications related to quantification of geophysical and biophysical parameters.

One of the most successful space programs dedicated to Earth observation is the American Landsat program. It consists of medium spatial resolution sensors with spectral bands strategically positioned in the electromagnetic spectrum to allow data acquisition ensuring both minimal redundancy and periodicity (temporal resolution).

The sensors onboard the Landsat satellite collect data in different regions of the electromagnetic spectrum by using different sensors throughout a program started in 1972 and which is continuing to this day: Multispectral Scanner System (MSS) and Return Beam Vidicon (RBV) were onboard the Landsat 1, Landsat 2, and Landsat 3 satellites; MSS and Thematic Mapper (TM) onboard the Landsat 4 and Landsat 5 satellites; Enhanced Thematic Mapper Plus (ETM+) onboard the Landsat 7 satellite; and Operational Land Imager (OLI) onboard the Landsat 8 satellite. These data are relayed to Earth receiving stations and processed into digitally encoded images stored on computers.

The digital images are originally represented by digital numbers. These numbers are defined in each element of spatial resolution (often called "pixel," which is a term originated from the words picture and element in English) according to the intensity of radiant electromagnetic energy flow (radiance) that focuses on a detector inside the sensor, which converts this intensity in an electrical signal that is converted to a digital number proportional to this electrical signal. Whereas this flow of electromagnetic energy comes from a portion of the Earth's surface with pre-established dimensions, the resulting digital number represents in fact a measure proportional to the "average" radiance of all objects inside this portion. Remember that this process occurs in specific spectral bands and independently. Depending on the sensor system and the altitude of the satellite, the spatial resolution of the image varies, since the dimensions of this portion of observed surface (pixel) vary. For example, the spatial resolutions of the MSS and TM sensors are approximately 0.45 hectares (57 m × 79 m) and 0.10 hectares (30 m × 30 m), respectively, on the Earth's surface.

An important phenomenon to take into consideration is the fact that the radiance that will lead to a digital number is a sum of integrated radiances from all objects or materials contained within the field of view (IFOV) snapshot of the sensor (which ultimately will lead to the pixel). Thus, the radiance effectively detected by the sensor will be explained by spectral mixture of various materials "in" pixel added to the atmospheric contribution. Therefore, the signal recorded by the sensor does not represent the physical and chemical composition of any object exclusively.

This phenomenon of "spectral mixture" has been considered by some researchers (e.g., Horwitz et al. 1971; Detchmendy and Pace 1972; Shimabukuro 1987). Usually the problem arises when trying to classify correctly a pixel containing a mixture of materials on the surface of the Earth such as soil, vegetation, rocks, water, and others. The nonuniformity of most natural scenes usually results in a large number of components in the mixture. The problem is further complicated by the fact that the proportion of specific materials contained "inside" a pixel can vary from pixel to pixel, generating varying degrees of ambiguity at the time of extracting information.

The spectral mixture becomes more critical in the application of digital image processing techniques than for the visual interpretation of images performed by trained interpreters using the so-called elements of interpretation, which are color, hue, texture, relative size, form, context, etc. The digital image classification is based predominantly on the radiometric/spectral characteristics of the pixels and then the classifier has the options: (1) classify the resolution element as a "pure" pixel when, in fact, it can contain only a small percentage of the "pure" material or (2) do not classify the pixel. The problem of spectral mixture is related to the problem of extracting "spectral signatures" or the spectral characterization of objects. To minimize problems caused by spectral mixture, it is necessary to have a better understanding of the effects of mixtures on pixel level.

This book presents the basic concepts that explain the spectral mixture, as well as the development of methods that aim to find the solution for different studies involving the application of remote sensing techniques. We will see that important aspects of these methods take into account the spectral characterization of different objects that compose the spectral mixture, and that based on this characterization it will be possible to quantify the proportions of each component contained inside a pixel. The solution of these methods is the fraction images, whose digital numbers represent the proportions (or percentages) corresponding to each objects in the mixture.

Thinking about environmental remote sensing or natural land resources, in general, the spectral mixture in a pixel is formed by basic components such as soil, vegetation, and shade. Then, after the solution of spectral mixture, we have the fraction images of soil, vegetation, and shade. These fraction images have been used in several areas of research such as those dedicated to forest resources, agriculture, urban studies, and assessment of flooded areas.

Fraction images have been used in several studies in different areas of application. In addition, these images have been used in projects to estimate deforested areas in the Brazilian Legal Amazon (PRODES), the detection of deforested areas in a near real time (DETER) also in Legal Amazon, and other projects such as the PANAMAZONIA II and AMAZONICA in estimating burned areas in the Amazon region. As we know, these projects analyze large areas on the ground through multispectral images with high temporal frequency. In that way, the fraction images that reduce the volume of data highlighting the information required for these projects are so important.

The purpose of this book is to offer to remote sensing users the opportunity to know the main aspects of spectral mixture, and the potential of linear spectral mixture model as a powerful tool for extracting information from remote sensing products.

São José dos Campos, SP, Brazil

Yosio Edemir Shimabukuro
Flávio Jorge Ponzoni

Contents

About the Authors

Yosio Edemir Shimabukuro is a Forest Engineer with a degree from the Federal Rural University of Rio de Janeiro, a Master's Degree in Remote Sensing from the National Institute for Space Research (INPE), and a PhD in Forest Sciences and Remote Sensing from Colorado State University (USA).

He works as a researcher in the Remote Sensing Division of INPE where, since 1973, he has been developing studies focused on the application of remote sensing techniques in the study of vegetation. He has been responsible for the design and improvement of methodologies for the identification and quantification of deforestation in the Amazon region, which have been applied in the PRODES and DETER projects developed by INPE. He is a permanent teacher of the postgraduate course in remote sensing of INPE.

Flávio Jorge Ponzoni is a Forest Engineer, with a Master's Degree in Forest Sciences from the Federal University of Viçosa and a PhD in Forest Sciences from the Federal University of Paraná. He completed a postdoctorate program at the Center for Meteorological and Climatic Research Applied to Agriculture at the State University of Campinas (CEPAGRI/UNICAMP), where he developed a work focused on the absolute calibration of orbital sensors.

He works as a researcher of the Remote Sensing Division of the National Institute for Space Research (INPE), where he is dedicated to studies of the spectral characterization of vegetation and the development of methodologies focused on the absolute calibration of orbital sensors. He also acts as a permanent teacher of the postgraduate course in remote sensing of INPE.

Chapter 1
Background

Abstract The main aspects related to the so-called spectral mixture under the perspective of orbital imagery carried out by Earth observation sensors are presented and contextualized.

Keywords Spectral mixture · Spatial resolution · Orbital imagery

Before further rationale for spectral mixture, it is necessary to understand the origins of all the concerns that have guided its concept in favor of the solution of several problems in the application of remote sensing techniques. It is necessary, for example, to know details about how effectively the intensity of electromagnetic radiation reflected by a particular portion of the Earth's surface is recorded by a sensor, but, before that, it will be addressed the origins of the so-called spectral mixture models.

Since the beginning of the application of remote sensing techniques, especially when the orbital images were available, according to some researchers, the use of multispectral data was limited, in part, by what they termed *mixture problem*, which occurs in the radiant flux reflected by a particular portion of the Earth's surface and that is instantly viewed by a sensor. Later the technical terms will be adequately defined and described that are applied to both the intensity of this flux and that portion instantly viewed by a sensor. However, for now, consider the following: a sensor is able to measure the intensity of radiant flux (of electromagnetic energy) of portions of the Earth's surface with defined dimensions. Typically these dimensions are considered "square" and represented by metric quantities, as 20 m × 20 m, 80 m × 80 m, 250 m × 250 m, and so on. Therefore, commonplace that portion instantly observed by one sensor is called "resolution element." When that resolution element is represented in an image, it is called "pixel," which is a word originated from the merging of the two words in the English language: picture and element.

Here we will assume the terms "resolution element" and "pixel" as synonymous, but the first term is related to the effective Earth's surface portion from which the

Y. E. Shimabukuro, F. J. Ponzoni, *Spectral Mixture for Remote Sensing*,
Springer Remote Sensing/Photogrammetry,
https://doi.org/10.1007/978-3-030-02017-0_1

intensity of the reflected electromagnetic energy flux is measured while the second one is the numerical representation of this intensity in an image.

The mixture is related with different materials or objects contained "inside" the resolution element (and "recorded" in the pixel) at the time of measuring the intensity of radiant flux by a sensor. The expression, "into the pixel," which often will be used in this book, assumes colloquial form, since, strictly speaking, there are no objects "inside" of a pixel. Actually, we are talking about a situation in which different objects are instantly viewed in an imaginary portion of the Earth's surface that has well-defined dimensions. The radiation flux originated by reflection of incident electromagnetic radiation is, in truth, a mixture of different fluxes of radiation that will result in a single measure of intensity in each spectral region in which the sensor is able to operate.

The concept of spectral mixture was discussed by Horwitz et al. (1971), Detchmendy and Pace (1972), Ranson (1975), and Heimes (1977), among others. The spectral mixture can occur in two cases:

- When the materials (or objects) are smaller than the resolution element size: In this case, the radiation flux detected by a sensor is composed by a mixture of different radiation fluxes reflected by different objects within the resolution element.
- When the materials (or objects) are greater than the resolution element size: In this case, the pixel overlaps the border between two or more materials or objects larger than its size.

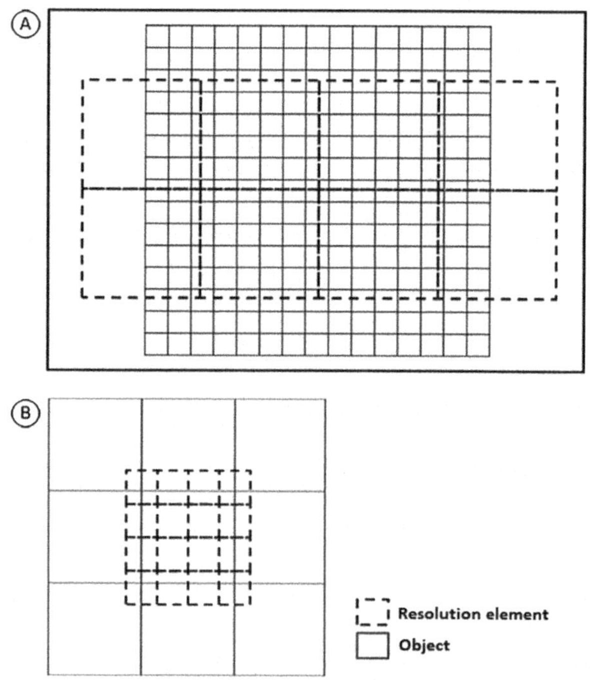

Fig. 1.1 Mixture problems: (**a**) caused by objects smaller than the resolution element (pixel) and (**b**) within the limits (boundaries) of the materials. (Source: Shimabukuro (1987))

In both cases, the signals recorded by the sensor are not representative of any of the objects present inside the resolution element. The idealized representation of mixture problem for both cases is illustrated in Fig. 1.1 that shows the objects scattered on the Earth's surface delineated by continuous lines and the pixels on the ground delineated by dashed lines.

Figure 1.2 shows schematically the mixture problem for the images generated by three sensors with different spatial resolutions and five classes (or objects) on the ground (a, b, c, d, and e). The sensor 3 presents the spatial resolution equal to *a*, the sensor 2 presents the spatial resolution equal to 2*a*, and the sensor 1 presents spatial resolution equal to 4*a* – that is, 10 m, 20 m, and 40 m, respectively. In that way, it can be seen that the sensor 1 (lowest spatial resolution) does not present any pixel with unique content (pure), the sensor 2 presents 5 pure pixels (1 of class b, 2 of class d, and 2 of class e), and the sensor 3 presents 36 pure pixels (8 of class a, 6 of class b, 12 of class d, 10 of class e, and 0 of class c).

The pixels painted in black color (Fig. 1.2) represent those occupied by only one (pure) of the classes on the ground, while the others refer to those that present mixture of classes with different proportions. Therefore the lower the sensor spatial

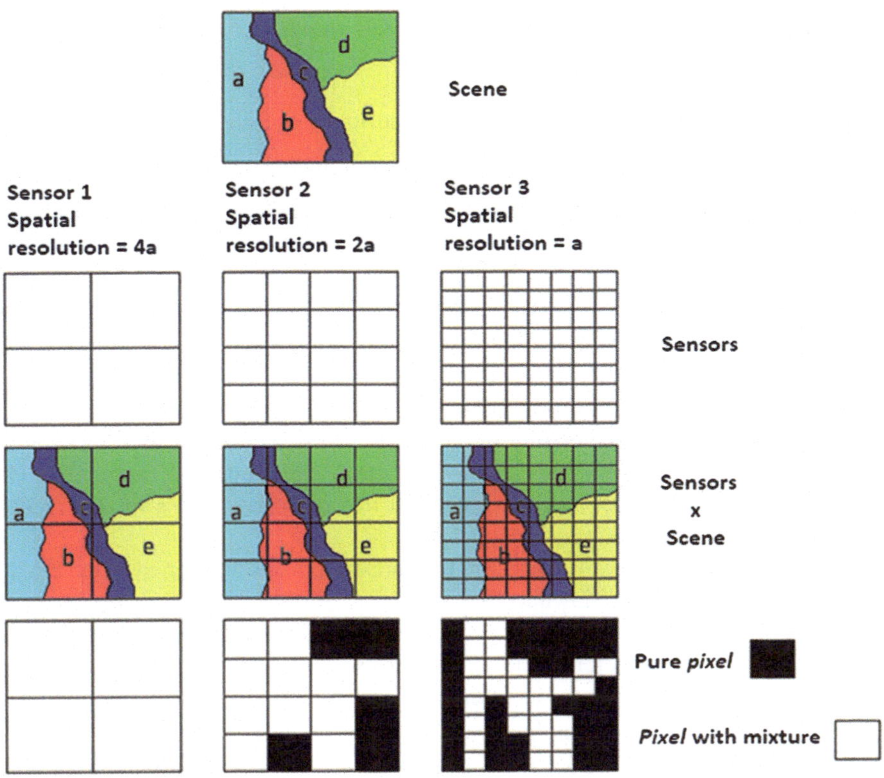

Fig. 1.2 Mixture for three sensors with different spatial resolutions and five classes (objects) on the ground. (Source: Piromal (2006))

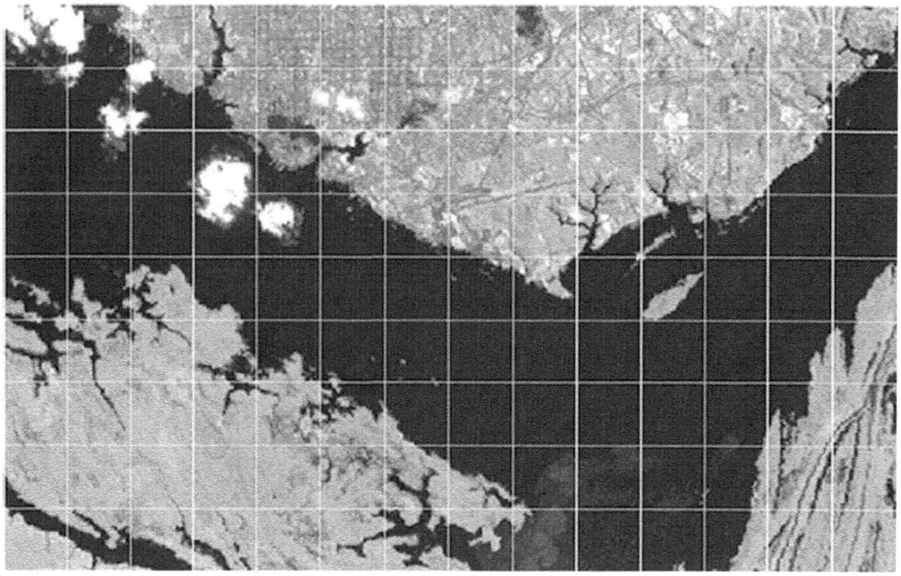

Fig. 1.3 TM/Landsat 5 (R5 G4 B3) image of the Manaus (AM) region and a grid corresponding to the size of the AVHRR pixels (1.1 km × 1.1 km)

resolution (i.e., the larger size of a pixel), the lower the chances of finding pure pixels in it.

Figure 1.3 shows an actual example for the region of Manaus (AM). Pixels of the AVHRR/NOAA (Advanced Very High Resolution Radiometer/National Oceanic and Atmospheric Administration) sensor (1.1 km × 1.1 km) represented by the grid over an image of the TM/Landsat 5 (Thematic Mapper on board of Landsat 5 satellite) sensor (30 m × 30 m) are composed of a mixture of water, soil, and vegetation. However, despite the coarser spatial resolution of the AVHRR sensor in relation to the spatial resolution of TM sensor, it is possible to find pure pixels of water due to the large extent of the Rio Negro.

Thus, the spectral characteristics of the sensor pixels such as the AVHRR/NOAA (approx. 120 ha), MODIS/Terra (6.25 ha), the MSS/Landsat 4 (approximately 0.45 ha), and TM/Landsat 5 (approximately 0.10 ha) on the Earth's surface can be affected by one or by both phenomena described earlier.

For better understanding the spectral mixture on a pixel, one can imagine a scatter plot similar to the one shown in Fig. 1.4.

In Fig. 1.4 the graph is composed, in *X*, by digital numbers or DN (whose origin will be described later) referring to an orbital image generated in the red spectral range and, in *Y*, the DNs generated in the near-infrared spectral range. For these two spectral bands and considering the spectral properties of the main natural resources (water, soil, and vegetation), the expected responses of these resources in this scatter plot would be the pixels occupied by water and/or shadow positioned closer to the origin of the graph, the pixels occupied by exposed soil positioned more distant to

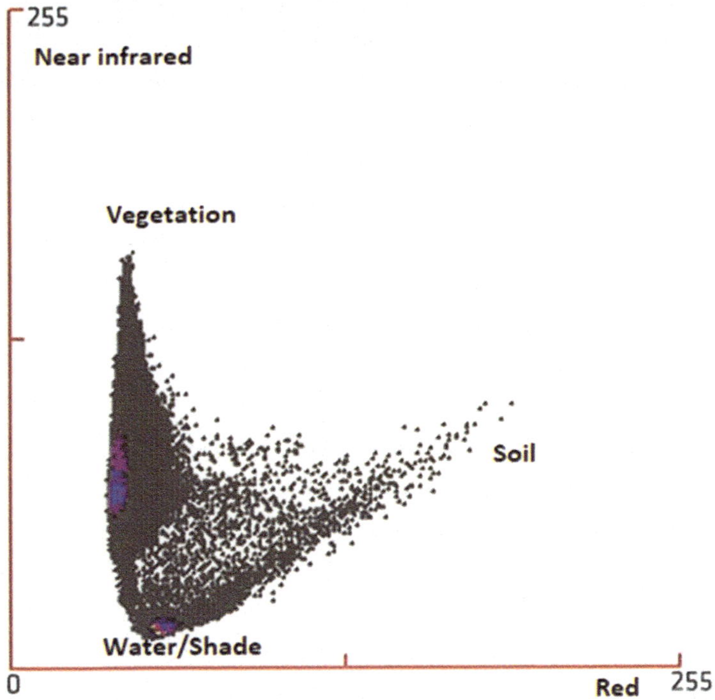

Fig. 1.4 Dispersion of the image pixels in the graph formed by the red and near-infrared spectral bands

the origin as *X* as well *Y*, and, finally, the pixels occupied by green vegetation positioned next to *Y*, but far from *X*.

Therefore, pure elements or pure objects, i.e., pixels fully occupied by only one of the natural resources considered, occupy pixels placed in the extreme of the triangular figure formed in this scatter plot. Thus, in the far superior vegetation occupies the entire pixels (in its whole area, or in 100%), while those placed on the far right bare soil occupies the entire pixels.

So, there is the question: of what are composed the pixels placed at the center of this triangular figure? The answer would be for equal proportions of the three natural resources, i.e., 33% of water, 33% of vegetation, and 33% of exposed soil. Considering "a walk" from the border of this triangular figure leaving for example, the vertex occupied by pure vegetation pixels, toward the origin of the scatter plot, gradually the pixels will gain proportions of water or shade in your composition until becoming the pure water or shade. Changing the direction of this path to the far right, in the place of these pure pixels of water or shade, gradually would be found pixels with higher proportions of soil until reaching pure pixels occupied by exposed soil. Similarly, we could consider that various proportions of these pure objects form the pixels within that triangular figure in the graph.

The mixture problem was first considered and addressed by Horwitz et al. (1971), who developed a technique to estimate the proportions of different types of land cover within a resolution element (or a pixel). However, it was not actually used in the remote sensing works until the middle of 1980 decade (Smith et al. 1985; Adams et al. 1986; Shimabukuro 1987). Since then, the interest of spectral mixture (linear and nonlinear) greatly increased, and several methods and applications have been developed for several areas of study (Boardman 1989; Roberts et al. 1993; Atkinson et al. 1997; Bastin 1997; Foody et al. 1997; Novo and Shimabukuro 1994; Shimabukuro et al. 1998; Rosin 2001; García-Haro et al. 2005; Alcântara et al. 2009).

This entire concept can be solved mathematically. For this, two approaches have been adopted:

- Least-squares approach
- Parameters estimation using a maximum likelihood approach

1.1 Mixture Problem in the Pixel

The first applications of mixture models aimed at solving the problem of area estimation by conventional methods of digital classification. Other way found in the literature to use spectral mixture models are the applications involving images derived from the proportions of the materials that compose the scene, which are the main focus of this book.

The problem of mixtures is very critical for the classification of varied themes related to land cover, limiting in any way the classification accuracy. Misclassification may occur when an area seen by a multispectral sensor contains two or more themes or classes of land cover, which produces a spectral response that does not match to the characteristics of any one of these classes (Ranson 1975).

Two general approaches have been taken to address the mixture problem:

- The technique of classification
- Attempts to model the relationships between the types and proportions of one class inside of a resolution element and the spectral response of this class (Heimes 1977)

In the first approach, two situations can occur:

- Classification of a pixel as a single class by using some decision function
- No classification of a pixel that do not have characteristic response of any one of individual class, i.e., leaving that pixel as unclassified (Heimes 1977)

The second modeling approach is more complex since it tries to explain the effects of the types and of the proportions of classes within a pixel related to its spectral characteristic. Pearson (1973) and Ranson (1975) presented the simple least squares approach and some considerations for practical applications. Ranson (1975) simulated the spectral characteristics or the spectral response of specific mixtures of different objects in order to reduce the effects of the mixture problem in digital classifications. Heimes (1977) evaluated the applicability of the least squares approach (Pace and Detchmendy 1973) using a set of well-defined data. In Heimes (1977) the observations and the proportions were obtained by simultaneous acquisition of radiometer data and photographic record of the scene.

Adams and Adams (1984) discussed the problem of separating the spectral responses of vegetation from rock/soil when these materials are present in a pixel. The goal was to extract information about rock/soils of pixels containing mixtures of rock/soils and vegetation. The approach used by the authors was based on the application of the linear model presented by Singer and McCord (1979). The authors concluded that the use of this model was successful in two Landsat scenes: one MSS image obtained in an area in Hawaii (USA) and another TM image obtained over the mountains of Tucson, Arizona (USA).

Adams et al. (1986) discussed the spectral mixture modeling applied to an image of the Viking Lander 1 sensor, which was one of the first sensors sent to Mars. The basic assumption was that the main factor of the spectral variation observed in the Viking 1 Lander image was the result of the linear mixture of materials present on the surface and in the shadow. If this hypothesis is valid, then a limited number of mixtures of spectra of the objects present in the scene (those that represent the principal constituents in the image) can determine all other spectra of other objects of the image, regardless of the instrumental calibration or the atmospheric effects.

Ustin et al. (1986), studying the applicability of data from TM/Landsat 5 sensor for the vegetation of semiarid region, used the spectral mixture model (Adams and Adams 1984; Adams et al. 1986), and they identified four spectral signatures of objects present in the scene. Those signatures were mixed in addictive mode to get the best fit in order to estimate the response of other objects, in pixel-to-pixel basis. The objects defined for their study were light soil, dark soil, vegetation, and shade representing the topographic variations.

When working with spectral mixture models, these objects of interest taken as a basis to estimate the responses of other objects are often called as "endmembers."

In the 1980 decade, during his PhD course in the United States, Shimabukuro (1987) developed and implemented the linear spectral mixture model applied to orbital data (Landsat MSS and TM).

At the end of 1980 decade and in the early 1990, with the advancement of technology in computer science, spectral mixture models began to be implemented in image processing systems, such as the image processing system (SITIM) and, subsequently, the georeferenced information processing system (SPRING), developed at the National Institute for Space Research (INPE). Later, similar models were becoming available in commercial image processing systems. With the availability of these models, large number of researchers and graduate students begun to explore this area of research.

In that way, the linear spectral mixture models have been used in various research works, and they are currently important tools to operationalize the large-scale projects, such as the estimation and the monitoring of deforested areas and burning areas in the Legal Amazon in digital approaches.

However, before to start studying details of these models, it is necessary to know relevant aspects of the origin of DNs present in the multispectral and hyperspectral images.

Chapter 2
The Origin of the Digital Numbers (DNs)

Abstract The origin of the digital numbers in orbital images is presented from both users' and engineer's point of views. This chapter aims to clarify important details of the generation of the digital numbers, which keep the spectral mixture, the central theme of this book.

Keywords Digital number · IFOV · EIFOV · Radiometry

The digital numbers (DNs) present in orbital images or even those generated by airborne sensors are generated according to a very simple principle: the electromagnetic radiation reflected from the Earth's surface and from the objects follows toward the sensor in a form of a flux with direction and intensity. That intensity is named radiance and can be measured in different spectral wavelengths. So, a flux of radiation reflected from the Earth's surface contains different "types" of electromagnetic radiation, differentiated themselves by wavelengths, which have their own strengths, or radiances. After the reflected radiation interact with the atmosphere during its trajectory toward the sensor, its intensity is measured in specific wavelength bands, according to the capacity of each sensor. These intensities are converted into electrical signals by specific detectors that register the radiation in specific spectral bands, and these electrical signals are converted into DNs by specific criteria in each spectral band.

The DNs are numerical values proportional to radiance (intensity) values measured in different spectral bands. The relationship with radiance is straightforward. The amplitudes of variation of the DNs are dependent on the number of bits adopted on generation of DNs. That number of bits in reality is the base 2 exponent, so that, if bits = 8, then $2^8 = 256$, i.e., the DNs will vary from 0 to 255 (256 levels of intensity or radiance). If bits = 10, so $2^{10} = 1024$, i.e., the DNs will vary from 0 to 1023 and so on. The amplitude of the DNs defines the radiometric resolution of the sensor.

But how is the relationship between the radiance effectively measured by the sensor and the corresponding DNs? These relationships are specific for each spectral band in which the sensor is designed to operate and are usually expressed by linear

Y. E. Shimabukuro, F. J. Ponzoni, *Spectral Mixture for Remote Sensing*, Springer Remote Sensing/Photogrammetry,
https://doi.org/10.1007/978-3-030-02017-0_2

equations. Equation 2.1 expresses a linear relationship between the radiance effectively measured by a sensor and DN:

$$\mathrm{DN}_\lambda = L_{0\lambda}.G_\lambda + \mathrm{offset}_\lambda \tag{2.1}$$

where:

DN_λ = DN value in spectral band λ
$L_{0\lambda}$ = radiance value effectively measured by the sensor in the spectral band λ
G_λ = angular coefficient of linear equation, also called gain, in spectral band λ
Offset_λ = linear equation intercept value, also called offset, in spectral band λ

Routinely $L_{0\lambda}$ receives the designation of apparent radiance or radiance at the top of the atmosphere. In international literature, usually published in the English language, this radiance receives the designation anachronistic TOA (top of atmosphere), so it is common to find the term TOA radiance.

Some authors, such as Chander et al. (2010), treat Eq. 2.1 differently. The treatment adopted by these authors is particularly interesting when those responsible for the distribution of sensors information perform it through the dissemination of what they call L_{min} and L_{max}. Many users are confused when presenting the regression line coefficients, which are nothing more than the absolute sensor calibration coefficients in each spectral band. L_{min} and L_{max} are thus the minimum and maximum radiances that the sensor is capable of measuring in a given spectral band. To make it easier to understand, please refer to Fig. 2.1.

In Fig. 2.1 there are two distinct ways to show the relationship between $L_{0\lambda}$ and DN_λ. In the chart of Fig. 2.1a, it has DN_λ as a function of $L_{0\lambda}$, which reflects the most physical point of view, that is, what is actually happening at the time of the acquisition of the data within the sensor. In the chart in Fig. 2.1b, it shows the point of view of the remote sensing users when DNs are used to calculate the corresponding $L_{0\lambda}$ values.

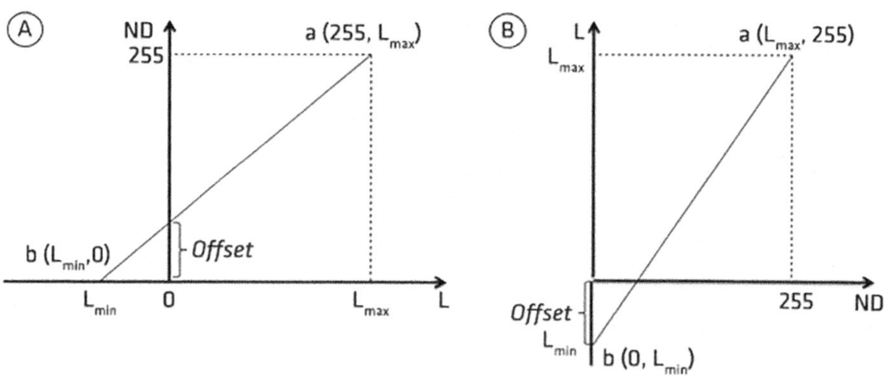

Fig. 2.1 Different forms of showing the relationship between $L_{0\lambda}$ and DN_λ

Note that in the relationship shown in the chart of Fig. 2.1a the offset values would be represented by amounts of DNs, whereas, in the ratio shown in the chart Fig. 2.1b the offset values would be represented by radiance units, which in this particular case would be negative. This explains the negative values of L_{min} presented by Chander et al. (2007) describing the relationship between $L_{0\lambda}$ and DN_{λ}. Often, people who are not familiar with the relationships described herein are wondering how it would be possible to find negative $L_{0\lambda}$ values. It is now understood that this is an algebraic peculiarity of the relationship between the two variables.

Chander et al. (2010), exploring the user's point of view (relationship between $L_{0\lambda}$ and ND_{λ} as shown in the chart in Fig. 2.1b), deduced the calculation of G_{λ} and $offset_{\lambda}$ as expressed in Eqs. 2.2 and 2.3.

$$G_{\lambda} = \frac{L_{\max\lambda} - L_{\min\lambda}}{DN_{\max\lambda} - DN_{\min\lambda}} \tag{2.2}$$

$$\text{offset}_{\lambda} = L_{\min\lambda} - \left(\frac{L_{\max\lambda} - L_{\min\lambda}}{DN_{\max\lambda} - DN_{\min\lambda}}\right) DN_{\min\lambda} \tag{2.3}$$

So

$$L_{0\lambda} = \left(\frac{L_{\max\lambda} - L_{\min\lambda}}{DN_{\max\lambda} - DN_{\min\lambda}}\right)\left(DN_{\lambda} - DN_{\min\lambda}\right) + L_{\min\lambda} \tag{2.4}$$

or

$$L_{0\lambda} = G_{\lambda}.DN_{\lambda} + \text{offset}_{\lambda} \tag{2.5}$$

where DN_{λ} refers to the DN value in the spectral band λ, which will be converted to apparent radiance $L_{0\lambda}$.

Gilabert et al. (1994) presented a didactic discussion about the various influent factors in $L_{0\lambda}$, which are represented schematically in Fig. 2.2.

In this figure, vectors represented by letter E refer to different intensities of electromagnetic incident radiation coming from a source. Taking into account that this intensity is called irradiance and that the source, in this case, is the sun (the main source of electromagnetic radiation exploited in remote sensing techniques applied to natural resources studies), we have $E_{0\lambda}$ representing the solar irradiance in the top of the atmosphere. This irradiance $E_{0\lambda}$ then begins its trajectory through the atmosphere toward the Earth's surface. Considering a specific target on that surface, this target actually receives both direct and diffuse fluxes of that incident radiation. Each one of these fluxes has its own intensities, i.e., its own irradiances. Thus, $E_{b\lambda}$ represents the incident irradiance directly on the target without interference of the atmosphere, and $E_{d\lambda}$ represents the diffuse incident irradiance on the target. This so-called diffuse radiation interacting with the atmosphere is scattered and hitting the target. These two fluxes, both direct and diffuse, interact with the target, and part of both is reflected toward space.

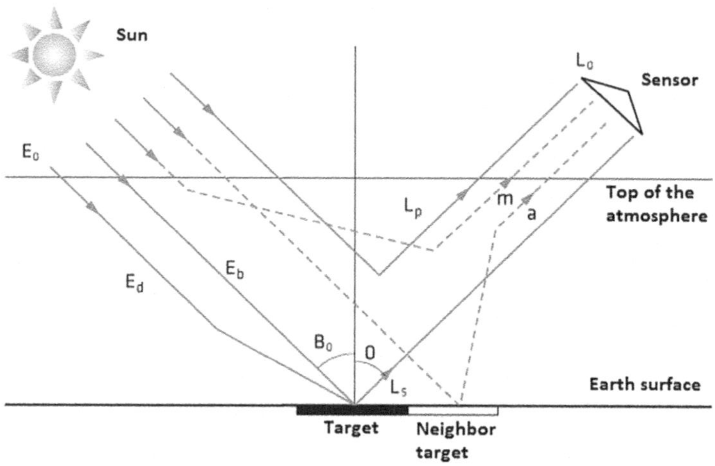

Fig. 2.2 Influent factors in $L_{0\lambda}$ (Source: Adapted from Gilabert et al. (1994))

Other fractions of $E_{0\lambda}$ interact with the atmosphere in direct and diffuse ways and do not even reach the target, having their trajectories altered back into space as represented by L_p and m. Besides that, other fractions of $E_{0\lambda}$ directly reach neighboring targets, and its reflection flux, in Fig. 2.2 represented by a, is also directed to space.

The DNs contained in orbital images or even generated by airborne sensors are then correlated to $L_{0\lambda}$ values, and therefore they cannot be directly associated with the spectral characteristics of the targets that we are intending to extract some information from. The spectral characterization of targets through the use of airborne or orbital images depends on the conversion of DNs into physical variables directly associated to the spectral properties of the targets, without the influence of the atmosphere and neighboring targets.

The conversion of $L_{0\lambda}$ in DN_λ is done in each spectral band in which the sensor was designed to operate. This means that the translation of $L_{0\lambda}$ to DN_λ is performed in a particular way and meeting specific criteria in each spectral band. Thus, the same DN value found in two or more images of different spectral bands may not represent the same value of $L_{0\lambda}$ effectively measured by the sensor. In this fictitious case, an object that should present differences in brightness levels in different spectral bands will appear with the same DN value, thus misleading its "spectral characterization."

Therefore, it is concluded that DN values from different spectral bands are not meant to characterize objects spectrally and neither to perform arithmetic operations between images of distinct spectral bands with the objective of associating the result with some geophysical or biophysical parameter of the specific target.

This does not mean that arithmetic operations between DNs of images generated in different spectral bands cannot be carried out. This depends on the goal the users want to achieve. When, for example, the users want to only visually enhance the objects or facilitate some sort of classification process (qualitative approach) that allows them to identify objects as they normally do in mapping jobs, such arithmetic operations are feasible. The problem arises when the interest is to explore the spectral differences of objects according to their spectral properties, since these will not be represented by DNs.

The values of G_λ, offset$_\lambda$, $L_{min\lambda}$, and $L_{max\lambda}$ are and should be made available by those responsible for generating or distributing the sensor data. This information is usually available in the form of metadata contained in specific files when accessing the images or when they are acquired in digital format. Some organizations inform these data on specific webpages. Of course the nomenclature adopted for these coefficients is quite variable, and it is up to the user to recognize it with caution and patience.

In the discussion presented by Gilabert et al. (1994) concerning the influent factors on the radiance $L_{0\lambda}$ illustrated in Fig. 2.2, other aspects related to the engineering adopted by the sensor were disregarded, which effectively registers the values of $L_{0\lambda}$. It is necessary to consider that the radiation flux that will originate $L_{0\lambda}$ and consequently its respective ND$_\lambda$ value is dependent on the spatial resolution of the sensor that is pre-established.

Figure 2.3 presents a schematic representation of the geometric components participants of the definition of spatial resolution of a sensor.

In Fig. 2.3, D refers to the dimensions of the detector, angular IFOV (Instantaneous Field of View) is the angular representation that defines the surface of which the value of $L_{0\lambda}$ is measured on the detector, F and H are respectively the focal length of the sensor and the flight altitude, linear IFOV is the linear dimension designed of the detector dimension D over the terrain, and EIFOV (Effective Instantaneous Field of View) is the effective dimension of the spatial resolution element from which the radiance $L_{0\lambda}$ is measured.

By the scheme presented in Fig. 2.3, it can be observed that the portion of the terrain from which the radiant flux originates whose $L_{0\lambda}$ intensity will be measured is often greater than that arising from the projection of D on the ground. It is worth saying that when one is working with data from a sensor whose nominal spatial resolu-

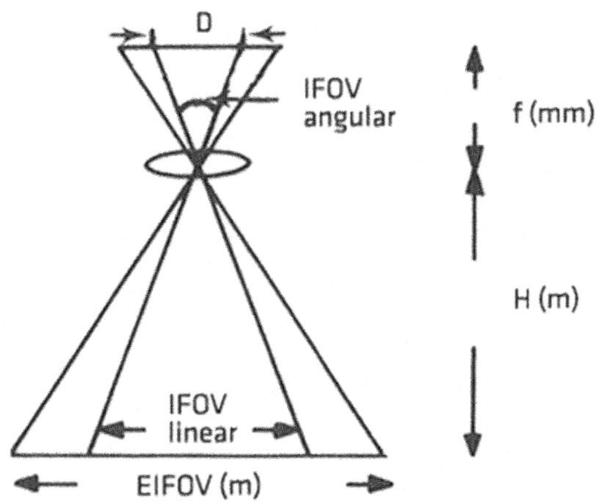

Fig. 2.3 Geometric components of the spatial resolution. (Source: Slater (1980))

tion is 30 m (30 m × 30 m in the *X* and *Y* directions), effectively it should be less than that, i.e., each spatial resolution element should have dimensions larger than 30 m.

Slater (1980) draws attention to this fact highlighting that one should not assume IFOV, EIFOV, and pixel terms as synonyms, since the latter is only an existing element in a numerical array in which a specific DN is placed.

The greater the differences between IFOV and EIFOV, it is understood that the system will be generating greater spatial distortions, which can be quantified by applying the modulation transfer function (MTF). More information about this application can be found at Slater (1980).

There are other definitions or appropriations of the term IFOV that vary somewhat from that recommended by Slater (1980). Other designations for the term and yet another term, FOV (Field of View), often associated with the angle of the optical aperture of a sensor or the cross track angle of imaging are presented in Fig. 2.4.

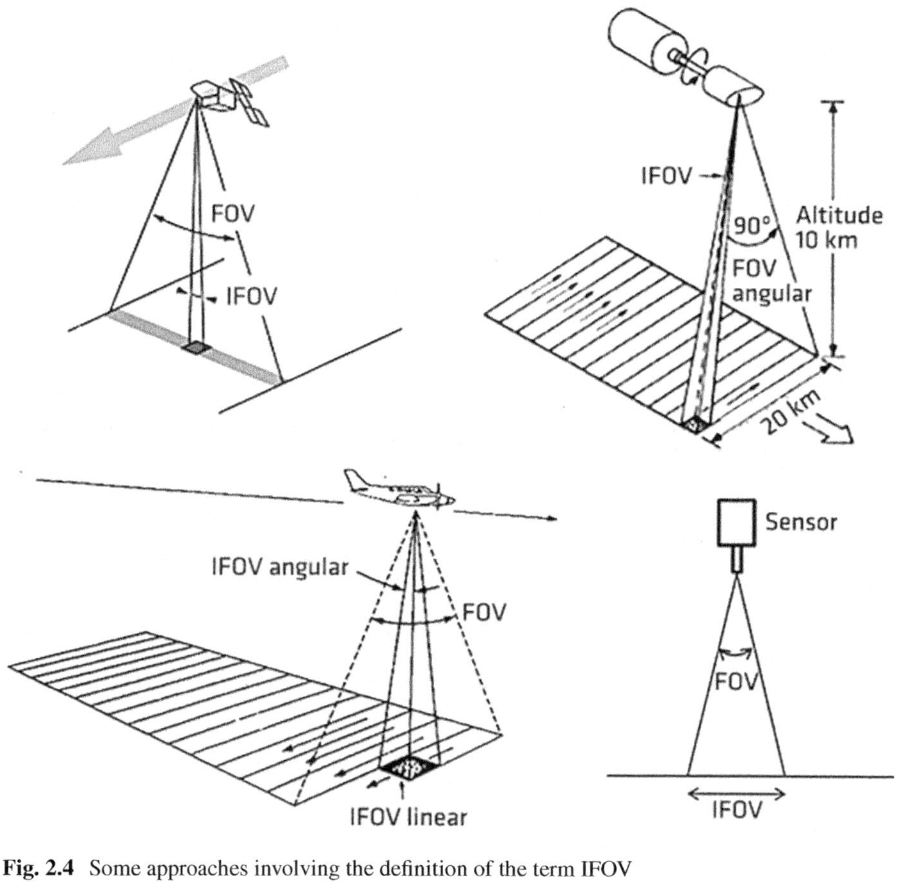

Fig. 2.4 Some approaches involving the definition of the term IFOV

It is worth noting that for the case of airborne or orbital sensors, the linear IFOV has dimensions in both *X* and *Y* axes since it is also often assumed the square form for the IFOV (or the EIFOV).

Whatever the size of the IFOV (or the EIFOV) and the way to define it (linearly or angularly), it must be taking into account that within it there are objects of different materials. The inclusion chance of objects with different spectral properties will be higher as large as the dimensions of IFOV (or EIFOV).

It should be noted that not only the proportion in the area of different objects within a resolution element on the ground is important, but its spectral properties is also important. It is important to remember that objects reflect, transmit, and absorb the incident electromagnetic radiation in a particular way, so it is easy to understand that each object contained within a resolution element on the ground will reflect more or less than another within a certain spectral band. Thus, considering a hypothetical situation in which within a resolution element there are four different objects occupying the same proportions, the value of radiance effectively measured in a given spectral band will be strongly influenced by the object that reflects more intensively in that same spectral band. This explains why the users can observe, in orbital images for example, objects whose dimensions are smaller than the IFOV of a sensor such as a road crossing a forest, for example. Normally such roads are very narrow and in theory should not be detectable by sensors whose IFOV has dimensions larger than the size of these roads. In this particular case, there is a very large radiometric contrast between the road reflectances (bare soil) and the vegetation cover in its border. The same could happen with a bridge and the water in its surroundings.

So, the radiance effectively measured by the sensor ends up being "contaminated" by the higher radiance generated by the road or by the bridge in relation to those generated by vegetation or water, respectively. This phenomenon is known as "spectral mixture," in which the spectral characteristic of a specific target prevails over the spatial resolution of a sensor. In these cases the reflected radiances of the road or of the bridge are greater than that of the forest and water, respectively, even occupying a smaller portion of the resolution element. Figure 2.5 presents didactic examples of such phenomenon.

In Fig. 2.5 tracings of a bridge and roads with dimensions lower than the IFOV can be observed in orbital images of different spectral regions. This phenomenon is very important when studying the mixing models because such proportions of spectral mixture are closely linked to it.

Important aspects of the origin of the DNs have been presented. They are present in sensor-generated images that feature specific characteristics from spectral, spatial, and temporal points of view. These features obviously offer opportunities

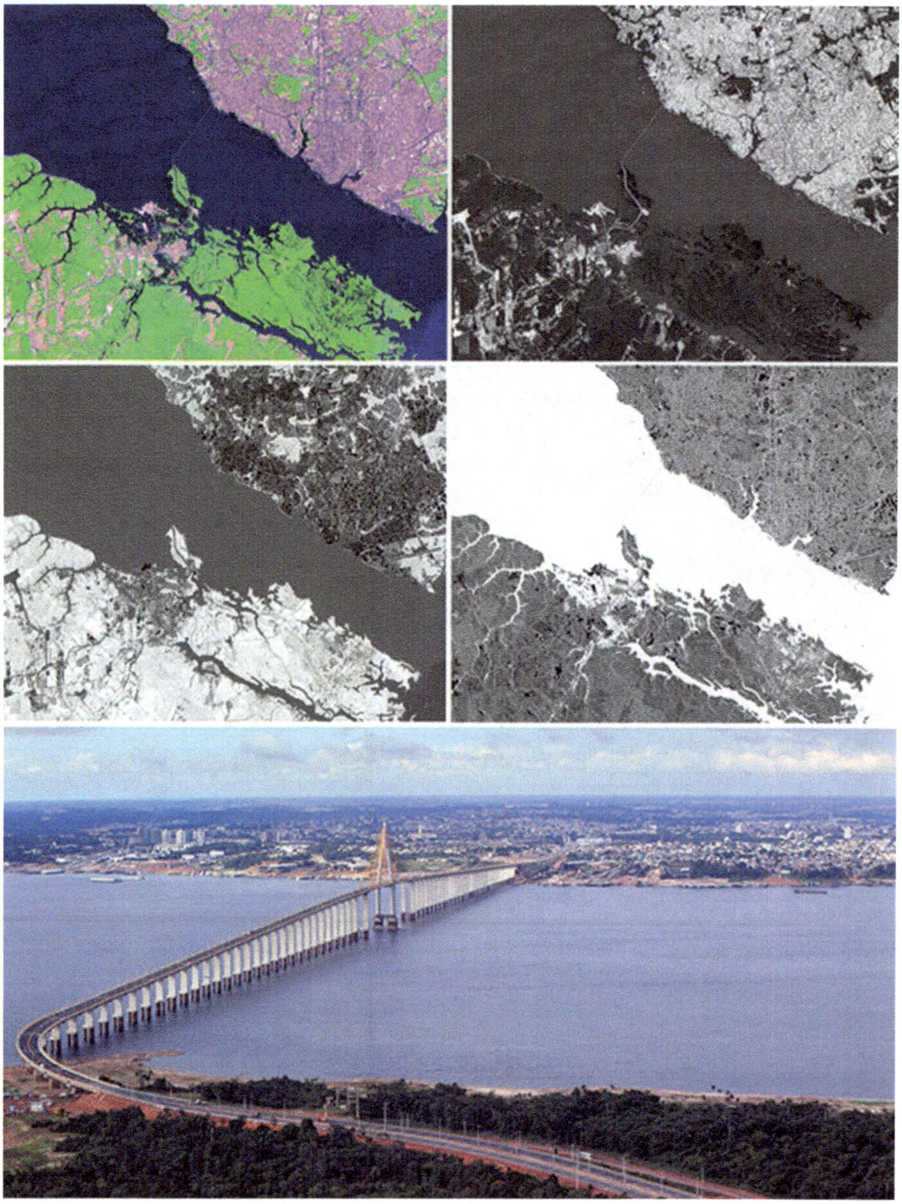

Fig. 2.5 Examples of possible object visualizations that present lower dimensions than those of the IFOV

and constraints to attend different applications. They also cannot be neglected when applying the linear spectral mixing model.

The following section presents some technical details of sensors whose data are often used when applying linear spectral mixing models.

Chapter 3
Orbital Sensors

Abstract The main technical characteristics of some more familiar sensors by the remote sensing community are presented in order to allow users to compare the possibilities of application of the linear spectral mixture models.

Keywords Earth observation sensors · Landsat · MODIS · HRV

The development of remote sensing techniques dates back to the end of the eighteenth century assuming that it started with the first photographic cameras and it had great momentum with the advent of the first space missions in the late 1950s and early 1960s. Throughout this process, numerous orbital sensors were developed for Earth observation that also initially aimed to generate data from the Earth's surface. So, through the data provided by sensors – such as those placed onboard satellites of the Landsat program (MSS, RBV, TM, ETM+, OLI) and the SPOT (High resolution Visible, HRV) program – it was attempted to develop thematic maps, which enabled monitoring the natural resources for decades.

Later, between the end of the 1980s and the early 1990s, the Earth Observing System (EOS) program arose in the United States, which pushed for the development and launching of sensors with the most varied capacities (resolutions). The objective was to generate data from the Earth's surface not only for mapping purposes but also for the quantification of geophysical and biophysical parameters, important in modeling studies and in varied forecasts, such as global climate change and future availability of natural resources.

Data from the most varied orbital sensors are currently available, some of which will be described in the following sections.

Y. E. Shimabukuro, F. J. Ponzoni, *Spectral Mixture for Remote Sensing*, Springer Remote Sensing/Photogrammetry,
https://doi.org/10.1007/978-3-030-02017-0_3

3.1 MODIS

The MODIS (Moderate Resolution Imaging Spectroradiometer) sensor was launched in the Earth's orbit by the United States in 1999 onboard the Terra Satellite (EOS AM) and, in 2002, onboard the Aqua satellite (EOS PM). This sensor acquires data in 36 spectral bands in the interval from 0.4 to 14.4 μm and in different spatial resolutions (2 bands with 250 m, 5 bands with 500 m, and 29 bands with 1 km). Together the instruments generate images from the whole Earth every day or 2 days. They are designed to provide large-scale measurements of the global dynamics of natural resources, including cloud cover changes, radiation balance, and processes occurring in the oceans, on the continents, and in the low atmosphere. MODIS was succeeded by the VIIRS (Visible Infrared Imaging Meter Suite) instrument onboard the satellite Suomi NPP (Suomi National Polar-orbiting Partnership), launched in 2011, and it will be replaced by future JPSS satellites (Joint Polar Satellite System).

With its low and moderate spatial resolution but high temporal resolution, MODIS data are useful for controlling landscape changes over time. Examples of such applications include the monitoring of vegetation health through time series analyses performed by vegetation indices (Lu et al. 2015), of the Earth's long-term cover changes to monitor deforestation rates (Klein et al. 2012; Leinenkugel et al. 2014; Lu et al. 2014; Gessner et al. 2015), of global snow cover trends (Dietz et al. 2012, 2013), and of flooding by virtue of precipitation, rivers, or floods in coastal areas due to the rise of sea level (Kuenzer et al. 2015) and the change of water levels of the Great Lakes (Klein et al. 2015), in addition to the detection and mapping of the forest fires in the United States. The United States Forest Service Remote Sensing Application Center (USFS) analyses MODIS images on a continuous basis in order to provide management and forest fire information.

The technical characteristics of this sensor are shown in Table 3.1.

The images from spectral bands 1 to 7 are commonly used in the Earth's surface studies. Spectral bands 3–7 are resampled from 500 to 250 m and can be used in the linear spectral mixing models. Spectral bands 21 and 22 are generated from the same wavelengths but present different saturation points.

3.2 SPOT Vegetation

The vegetation program is jointly developed by France, the European Commission, Belgium, Italy, and Sweden. The first satellite of the program, vegetation 1, was launched on 24 March 1998, onboard the SPOT 4 satellite, while the second instrument, vegetation 2, was released on 4 May 2002 onboard the SPOT 5 satellite. They provide data to monitor the Earth's surface parameters with daily frequency on a global basis with an average spatial resolution of 1 km. The ground segment associated with the program processes the data generated to provide standard products to

Table 3.1 Technical characteristics of the MODIS sensor

Spectral bands	Wavelength (nm)	Spatial resolution (m)	Usage
1	620–670	250	Earth/cloud/aerosol limits
2	841–876	250	
3	459–479	500	Earth/cloud/aerosol properties
4	545–565	500	
5	1230–1250	500	
6	1628–1652	500	
7	2105–2155	500	
8	405–420	1000	Ocean color/phytoplankton/ biogeochemistry
9	438–448	1000	
10	483–493	1000	
11	526–536	1000	
12	546–556	1000	
13	662–672	1000	
14	673–683	1000	
15	743–753	1000	
16	862–877	1000	
17	890–920	1000	Atmospheric water vapor
18	931–941	1000	
19	915–965	1000	
20	3.660–3.840	1000	Cloud/surface temperature
21	3.929–3.989	1000	
22	3.929–3.989	1000	
23	4.020–4.080	1000	
24	4.433–4.498	1000	Atmospheric temperature
25	4.482–4.549	1000	
26	1.360–1.390	1000	Cirrus water vapor
27	6.535–6.895	1000	
28	7.175–7.475	1000	
29	8.400–8.700	1000	Cloud properties
30	9.580–9.880	1000	Ozoniun
31	10.780–11.280	1000	Cloud/surface temperature
32	11.770–12.270	1000	
33	13.185–13.485	1000	Cloud altitude
34	13.485–13.785	1000	
35	13.785–14.085	1000	
36	14.085–14.385	1000	

the user community. The entire system complements the high-resolution spatial capacity available on the sensors of the SPOT series (*Satellite Pour la Observation de la Terre*) providing similar and simultaneous spectral measurements in the visible and infrared spectral regions. The original features of the instruments allow users to have access to:

(a) Robust and simple multitemporal measurements of solar radiant properties, continuous and global monitoring of continental areas, data generation for regional or local studies, extensive data set with accurate calibration and localization, and continuity and consistency, which will be made available by future generations of these sensors
(b) Multiscale approaches

The decision to carry out this program was the result of the development of many studies and projects during the last 20 years: the use of remote sensing data in operational programs or projects that should drive operating applications has increased strongly until the availability and quality of the data became clearly a limitation.

As working groups, user communities and international programs have been expressing their needs to increase the details in different domains (spectral, radiometric, temporal, and spatial). The idea of taking the opportunity to embarking on a dedicated and definitely operational mission onboard SPOT 4 was supported by the program's partners. The needs of the European Commission's sectoral policies for the management of production in agriculture, for forestry, for environmental monitoring, and for national partners, as well as for major international programs related to the study of global changes, have been synthesized by an international committee of users and have become the basis for the technical development of the entire system.

The strong commitment of the European Commission was also a clear sign that the mechanisms by which remote sensing systems are designed and used are changing. Taking into account that the methodologies for using remote sensing data become more adapted to a regular and operational need, the decision to undertake such development is now also in the hands of users and is not only under exclusive responsibility of the space agencies. The development of vegetation sensors, the structure of the program, and its achievements constitute a test by which the new mechanisms are exemplified. Their overall goals, however, should remain as a long-term commitment to provide useful data for the user community.

Table 3.2 presents the main technical characteristics of the vegetation 1 and vegetation 2 sensors.

Table 3.2 Technical characteristics of the vegetation 1 and vegetation 2 sensors

Sensor	Spectral band	Spectral resolution	Spatial resolution	Temporal resolution	SWATH
Vegetation 1 and Vegetation 2	B0	0.43–0.47 μm	1.15 km	24 h	2,250 km
	B2	0.61–0.68 μm			
	B3	0.78–0.89 μm			
	SWIR	1.58–1.75 μm			

3.2.1 *Landsat MSS, TM, ETM+, and OLI*

The Landsat satellite family began with the launch of Landsat 1 in 1972, which carried onboard the one that would be the first successful Earth observation sensor: the so-called Multispectral Scanner System (MSS). The Thematic Mapper (TM) sensor began to be used 10 years later when, in 1982, it was placed onboard the Landsat 4 satellite.

The longest satellite of the program Landsat was the Landsat 5, also carrying onboard the TM sensor, which was launched in 1984 and was discontinued in 2013. This sensor generated images of excellent quality up to 2011, something unprecedented until then in terms of time period in orbit. The Landsat 7 satellite was launched in 1999, taking the enhanced Thematic Mapper Plus (ETM+) sensor onboard with the increase of a panchromatic band compared to its predecessor, the TM Landsat 5.

The Landsat 8 brought innovations with the replacement of the TM sensor by the Operational Land Imager (OLI) sensor characterized by the narrowing of spectral bands and the inclusion of the coastal/aerosol, thermal, and cirrus bands.

Table 3.3 presents the characteristics of the TM, ETM+, and OLI sensors of the Landsat program.

The radiometric resolution of the TM and ETM+ sensors is 8 bits, while the sensor OLI is 12 bits. Their temporal resolution is 16 days.

3.3 Hyperion

The Hyperion sensor operates in 220 spectral bands between 0.4 and 2.5 μm, which is why it is called hyperspectral. Launched in November 2000 as part of the EOS program, it had as main objective to initiate the availability of a series of innovative Earth observation sensors. The innovation was then focused on the generation of

Table 3.3 Technical characteristics of the TM, ETM+, and OLI sensors

Spectral bands	TM (μm)	ETM+(μm)	OLI (μm)	Spatial resolution
Coast/aerosol			0.433–0.453	30 m
Blue	0.45–0.52	0.45–0.52	0.450–0.515	30 m
Green	0.53–0.61	0.53–0.61	0.525–0.600	30 m
Red	0.63–0.69	0.63–0.69	0.630–0.680	30 m
Near infrared	0.78–0.90	0.78–0.90	0.845–0.885	30 m
Shortwave infrared	1.55–1.75	1.55–1.75	1.560–1.660	30 m
Thermal	10.4–12.5	10.4–12.5		120 m/60 m
Shortwave infrared	2.09–2.35	2.09–2.35	2.100–2.300	30 m
Panchromatic		0.52–0.90	1.360–1.390	15 m
Cirrus			0.52–0.90	30 m

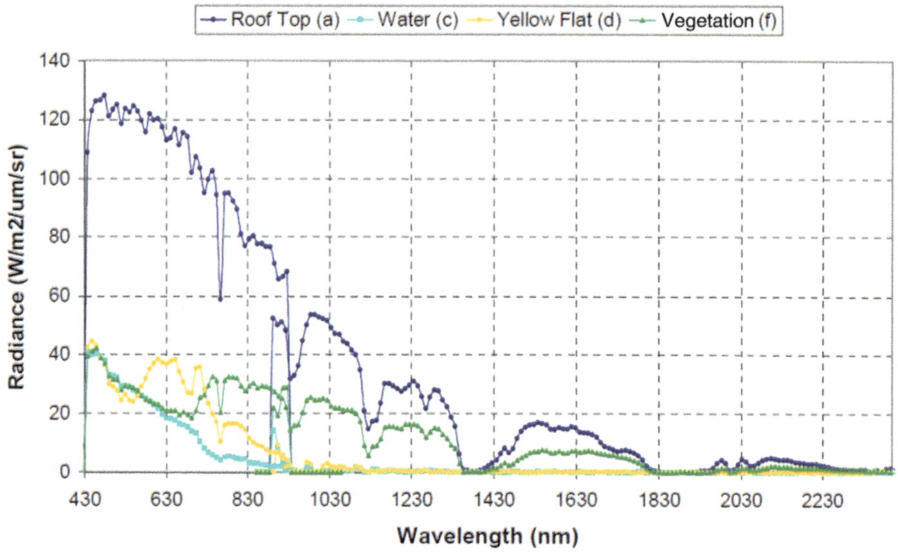

Fig. 3.1 Example of spectra that can be generated based on the Hyperion sensor data

spectral data that allowed the quantification of geophysical and biophysical parameters through the spectral characterization of the objects.

This sensor was placed onboard the satellite EO-1, which has sun-synchronous orbit at altitude of 705 km, and it has an imaging push broom sensor with a width of 7.65 km, spatial resolution of 30 m, and a 12-bit radiometric resolution.

Ideally the spectral characterization of objects intended by the analysis of data generated by this sensor should happen in pure pixels, that is, in those pixels where the spectral mixture does not occur. However, it is known that even for miniature dimensions of IFOV, there will always be the spectral mixture as a practically mandatory phenomenon.

Figure 3.1 illustrates an example of spectra that can be generated based on the Hyperion sensor data.

From what was previously exposed, it is clear that spectra such as those observed in Fig. 3.1 do not refer to pure objects but rather to mixtures of objects contained in pixels of 30 m × 30 m.

The conception of new sensors is a constant. There are countless innovations planned for the next decades in the different areas explored by remote sensing techniques (spatial, spectral, radiometric, and temporal). Regardless of innovations it is certain that the remote sensing data-user community will have at its disposal an immense amount of the Earth's surface data. In addition, spectral mixing models also serve as an alternative to reducing the volume of data to be processed aiming at the information extraction process.

Chapter 4
The Linear Spectral Mixture Model

Abstract The linear spectral mixture model is presented in its math concept. During this discussion, examples are presented in order to facilitate the reader's understanding of the concepts involved in the design of the model.

Keywords Linear spectral mixture model · LSMM concept · Spectral mixture modeling

The time has come to understand a bit more how it is possible to calculate the fractions or proportions of each "object" within a pixel, which can assume various dimensions depending on the spatial resolution of each sensor.

These fractions are calculated by applying mathematical models. This chapter will deal with linear models, but it is important to know that the linearity may not be the only way to describe the contribution of each object within a pixel. The linear spectral mixture models (LSMM) will be described because they have been widely used by researchers with consistent results.

Following this approach, the spectral response in each pixel, in any spectral band of the sensor, can be considered as a linear combination of the spectral responses of each component present in the mixture. Thus, each pixel in the image, which can assume any value within the gray level range (2^n bits), contains information about the proportion (amount) and the spectral response of each component within the resolution unit on the ground. Therefore, for any multispectral image generated by any sensor system, considering the knowledge of the components proportion, it will be possible to estimate the spectral response of each one of these components.

Similarly, if this response is known, then the proportion of each component in the mixture can be estimated. This characteristic will help in the analysis of different sensors with diverse spatial resolution. For example, it is possible to generate the fraction images from a sensor with high spatial resolution (pixels with small dimensions) and, based on these proportions, to estimate the spectral responses of the objects present in pixels generated by a medium spatial resolution sensor, and then to generate the fraction images for this sensor images (Shimabukuro and Smith 1995).

Y. E. Shimabukuro, F. J. Ponzoni, *Spectral Mixture for Remote Sensing*, Springer Remote Sensing/Photogrammetry,
https://doi.org/10.1007/978-3-030-02017-0_4

Imagine a situation as simple as possible in which a panchromatic image is available, obtained in a relatively wide spectral range, generated from a sensor of 8-bit radiometric resolution, i.e., with 256 gray levels. In this case, the system of equations of the linear spectral mixture model can be formulated as follows:

$$R = bx_1 + px_2 \tag{4.1}$$

$$x_1 + x_2 = 1 (\text{sum of proportions must be equal to } 1) \tag{4.2}$$

where:

R = spectral response of the pixel of the image
b = spectral response of the bright object
p = spectral response of dark object
x_1 = proportion of the bright object
x_2 = proportion of dark object

In this way, it would be possible to generate two fraction images (bright and dark) for this panchromatic image. The fraction images would be the solution of the system of equations for all image pixels, as shown below.

Making:

$$x_2 = 1 - x_1 \tag{4.3}$$

Replacing then Eq. 4.3 in Eq. 4.1,

$$R = bx_1 + p(1 - x_1) = bx_1 + p - px_1 = x_1(b - p) + p \quad or \quad R - p = x_1(b - p) \tag{4.4}$$

As the image DNs range from 0 to 255, it can be considered that there is a bright pure pixel (b = 255) and other dark pure pixel (p = 0). In this case, the pixels with values between 1 and 254 would be a mixture of responses of these pixels. So, it would be possible to generate two fraction images (bright and dark) for this panchromatic image. The fraction images would be the result of the solution of Eq. 4.4 for all pixels of the image.

For example:

For $R = 0$, substituting the values of the pixels (b = 255) and dark (p = 0) in Eq. 4.4, it is obtained $0 - 0 = x_1 (255 - 0) = 255x_1$ and $x_1 = 0$ and, by Eq. 4.2, $x_2 = 1 - 0 = 1$ (dark pixel).

For $R = 255$, substituting the values of the pixels (b = 255) and dark (p = 0) in Eq. 4.4, it is obtained $255 - 0 = x_1 (255 - 0) = 255x_1$ and $x_1 = 1$ and, by Eq. 4.2, $x_2 = 1 - 1 = 0$ (bright pixel).

For $R = 127$, substituting the values of the pixels (b = 255) and dark (p = 0) in Eq. 4.4, it is obtained $127 - 0 = x_1 (255 - 0) = 255x_1$ and x_1 = approximately 0.5 and, by Eq. 4.2, $x_2 = 1 - 0.5 = 0.5$ (mixture pixel).

Making this for all the image pixels, the bright and dark fraction images were obtained as the example for a real image shown in Fig. 4.1.

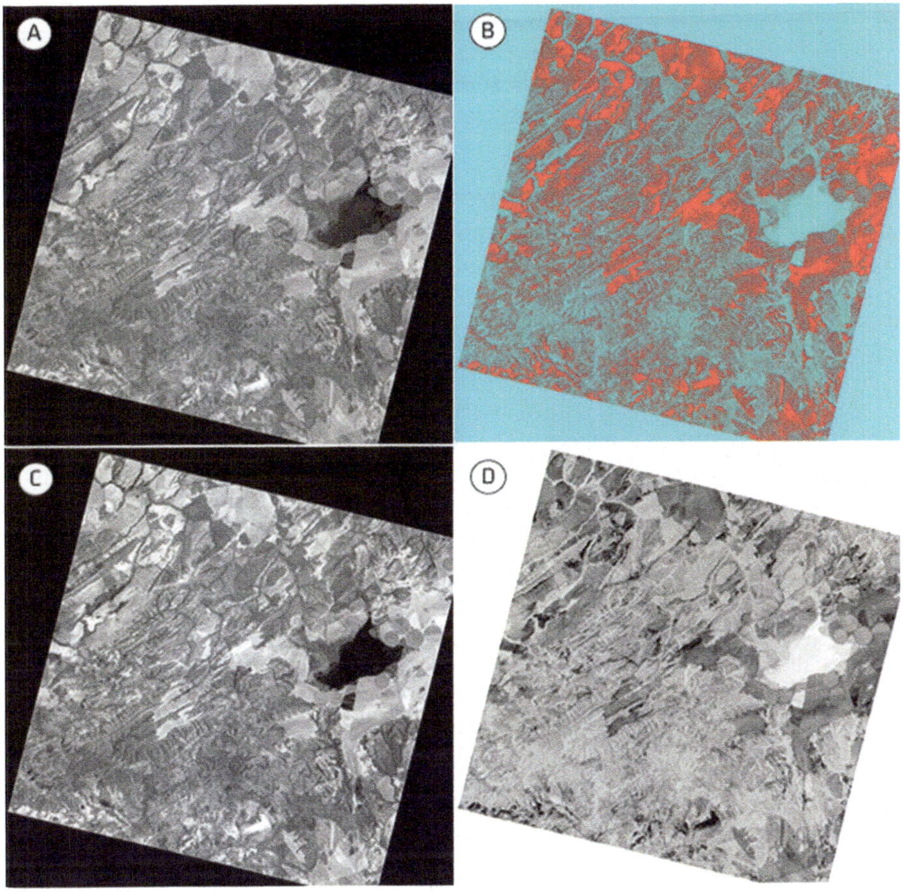

Fig.4.1 (**a**) HRC CBERS panchromatic image; (**b**) Fraction image color composite; (**c**) bright fraction image; (**d**) dark fraction image

The resulting fraction images could be used together or individually, depending on the objectives to be achieved. Although it has not been used normally, the linear mixing model for two endmembers serves to show that the solution of this mixture model is not so complex as will be shown in case of three or more endmembers in the mixture.

Generally, then, the linear spectral mixture model can be written as:

$$
\begin{array}{lllllll}
R_1 & = & a_{11}x_1 + & a_{12}x_2 + & \cdots & +a_{1n}x_n & +e_1 \\
R_2 & = & a_{21}x_1 + & a_{22}x_2 + & \cdots & +a_{2n}x_n & +e_2 \\
\vdots & & \vdots & \vdots & & \vdots & \vdots \\
R_m & = & a_{m1}x_1 + & a_{m2}x_2 + & \cdots & +a_{mn}x_n & +e_m
\end{array}
$$

or

$$R_i = \text{sum}\left(a_{ij}x_j\right) + e_i \tag{4.5}$$

where:

R_i = average spectral reflectance for the i-th spectral band
a_{ij} = spectral reflectance of the j-th component in the pixel for the i-th spectral band
x_j = value of the j-th component in the pixel
e_i = error for the i-th spectral band
$j = 1, 2, ..., n$ (n = number of components undertaken for the problem)
$i = 1, 2, ..., m$ (m = number of spectral bands for the sensor system)

As mentioned previously, this model assumes that the spectral response (in Eq. 4.5, expressed as reflectance) of the pixels are linear combinations of the spectral response of the components within the pixel. To solve Eq. 4.5, it is necessary to have the spectral reflectance of the pixels in each band (R_i) and the spectral reflectance of each component (a_{ij}) in each band in order to estimate the proportion values, or vice versa.

The linear spectral mixture model is a typical example of inversion problem (indirect measures) in remote sensing. Some concepts of inversion problem and three mathematical approaches to the solution of the system of linear equations will be discussed below.

On the inversion problem, the mixing model without the relative term to the error (e_i), which was previously defined, can be rewritten in the matrix form:

$$R = Ax \tag{4.6}$$

where:

A = matrix of m rows by columns containing input data, representing the spectral reflectance of each component
R = vector of m columns, representing the pixel reflectance
x = vector of n columns, representing the proportion values of each component in the mixture (variables to be estimated)

The procedure to solve a problem of remote sensing such as of Eq. 4.6 is called the inversion problem or method of indirect measures. In this case, the average spectral reflectance of the pixel (R) is assumed to be linear dependent of spectral reflectance of each component (A). Therefore, the proportion value (x_j) will be zero if the respective a_{ij} and R_i are not dependent on each other.

Numerical inversions can produce results that are mathematically correct, but physically unacceptable. It is important to understand that most of the problems of physical inversion are ambiguous, since they do not have a unique solution and a discreet reasonable solution is achieved by imposing additional boundary conditions.

In remote sensing, the users are typically interested in knowing the state of a physical quantity, biological or geographic (or several of them), such as the biomass of a specific agricultural crop, the amount of a pollutant gas into the atmosphere or the extent and the condition of the global coverage of snow in a given date.

For the solution of system of linear equations that represents the spectral mixture model, there are several mathematical approaches based on the method of least squares. Following it will be presented three algorithms that are available in the current image processing software (SPRING, ENVI, PCI).

4.1 Mathematical Algorithms

As seen previously, the linear spectral mixture model is a system of equations, with one equation for each considered sensor band. For example, for the MSS, there are four equations, corresponding to the bands 4, 5, 6, and 7, while for the TM, there are six equations, corresponding to the bands 1, 2, 3, 4, 5, and 7, taking into account only the solar optical spectrum. It is important to keep in mind that it is not necessary to use all the available bands, but it must obey the condition that the number of reference spectra (or pure pixels) always be less than the number of spectral bands. In this way, mathematical algorithms are required for the solution of the system of equations formed by the spectral response of the pixel, which is the function of the proportion of each reference spectrum (or pure pixels) weighted by the respective spectral response of the endmember.

These reference spectra or pure pixels are routinely referred as endmembers, a designation that the users more familiar with applying linear mixture models are more accustomed. So from this point, it will be adopted that name.

Following three mathematical algorithms will be presented: constrained least squares (CLS), weighted least squares (WLS), and principal components (PC).

4.1.1 Constrained Least Squares (CLS)

This method estimates the proportion of each component within the pixel minimizing the sum of squared errors. The proportion values must be nonnegative (physical meaning) and add to 1. To solve this problem, it was developed a method of quasi-closed solution (e.g., a method which finds the solution making approximations that satisfy these constraints). In this case, the proposed method will be presented for the cases of three or four components within the pixel. It is important to keep in mind that this model can be developed for a greater number of endmembers but the solution is becoming increasingly complex, as it will be seen between three and four endmembers models. Thus, the mixture model can be written as:

$$
\begin{array}{ccccc}
r_i & = & a_{11}x_1 + & a_{12}x_2 + & a_{13}x_3 + & e_1 \\
\vdots & & \vdots & \vdots & \vdots & \vdots \\
r_m & = & a_{m1}x_1 + & a_{m2}x_2 + & a_{m3}x_3 + & e_m
\end{array}
$$

It is possible to write as:

$$r_i = \Sigma\left(a_{ij}x_j\right) + e_i \tag{4.7}$$

or,

$$e_i = r_i - \Sigma\left(a_{ij}x_j\right) \tag{4.8}$$

The function to be minimized is:

$$F = \Sigma e_i^2 \tag{4.9}$$

where m is the number of sensor spectral bands used, for example, $m = 4$ for the MSS sensor or $m = 6$ for the TM sensor.

4.1.2 Four Spectral Bands and Three Components

In this case, the mixture problem can be written as:

$$r_1 = a_{11}x_1 + a_{12}x_2 + a_{13}x_3 + e_1$$

$$r_2 = a_{21}x_1 + a_{22}x_2 + a_{23}x_3 + e_2$$

$$r_3 = a_{31}x_1 + a_{32}x_2 + a_{33}x_3 + e_3$$

$$r_4 = a_{41}x_1 + a_{42}x_2 + a_{43}x_3 + e_4$$

The function to be minimized is:

$$\begin{aligned} e_1^2 + e_2^2 + e_3^2 + e_4^2 = E_1x_1^2 + E_2x_2^2 + E_3x_3^2 + E_4x_1x_2 + E_5x_1x_3 \\ + E_6x_2x_3 + E_7x_1 + E_8x_2 + E_9x_3 + E_{10} \end{aligned} \tag{4.10}$$

The values of the coefficients E_1 to E_{10} are shown in Table 4.1.

Considering the first restriction, $x_1 + x_2 + x_3 = 1$ or $x_3 = 1 - x_1 - x_2$, and replacing this restriction in Eq. 4.10, the function to be minimized becomes:

$$e_1^2 + e_2^2 + e_3^2 + e_4^2 = A_1x_1^2 + A_2x_2^2 + A_3x_1x_2 + A_4x_1 + A_5x_2 + A_6 \tag{4.11}$$

Table 4.1 values of the coefficients *E* of Eq. 4.10

$E_1 = a_{11}^2 + a_{21}^2 + a_{31}^2 + a_{41}^2$
$E_2 = a_{12}^2 + a_{22}^2 + a_{32}^2 + a_{42}^2$
$E_3 = a_{13}^2 + a_{23}^2 + a_{33}^2 + a_{43}^2$
$E_4 = 2 \times (a_{11}\, a_{12} + a_{21}\, a_{22} + a_{31}\, a_{32} + a_{41}\, a_{42})$
$E_5 = 2 \times (a_{11}\, a_{13} + a_{21}\, a_{23} + a_{31}\, a_{33} + a_{41}\, a_{43})$
$E_6 = 2 \times (a_{12}\, a_{13} + a_{22}\, a_{23} + a_{32}\, a_{33} + a_{42}\, a_{43})$
$E_7 = -\, 2 \times (a_{11}\, r_1 + a_{21}\, r_2 + a_{31}\, r_3 + a_{41}\, r_4)$
$E_8 = -\, 2 \times (a_{12}\, r_1 + a_{22}\, r_2 + a_{32}\, r_3 + a_{42}\, r_4)$
$E_9 = -\, 2 \times (a_{13}\, r_1 + a_{23}\, r_2 + a_{33}\, r_3 + a_{43}\, r_4)$
$E_{10} = r_1^2 + r_2^2 + r_3^2 + r_4^2$

Table 4.2 Coefficients values of *A* of Eq. 4.11

$A_1 = a_{11}^2 + a_{21}^2 + a_{31}^2 + a_{41}^2 + a_{13}^2 + a_{23}^2 + a_{33}^2 + a_{43}^2 - 2\times (a_{11}\, a_{13} + a_{21}\, a_{23} + a_{31}\, a_{33} + a_{41}\, a_{43})$
$A_2 = a_{12}^2 + a_{22}^2 + a_{32}^2 + a_{42}^2 + a_{13}^2 + a_{23}^2 + a_{33}^2 + a_{43}^2 - 2\times (a_{12}\, a_{13} + a_{22}\, a_{23} + a_{32}\, a_{33} + a_{42}\, a_{43})$
$A_3 = 2\, (r_1^2\, a_{13}^2 + a_{23}^2 + a_{33}^2 + a_{43}^2 + a_{11}\, a_{12} + a_{21}\, a_{22} + a_{31}\, a_{32} + a_{41}\, a_{42} - a_{11}\, a_{13} - a_{21}\, a_{23} - a_{31}\, a_{32} - a_{41}\, a_{43} - a_{12}\, a_{13} - a_{22}\, a_{23} - a_{32}\, a_{33} - a_{42}\, a_{43})$
$A_4 = 2\, (-\, a_{13}^2 - a_{23}^2 - a_{33}^2 - a_{43}^2 + a_{11}\, a_{13} + a_{21}\, a_{23} + a_{31}\, a_{33} + a_{41}\, a_{43} - a_{11}\, r_1 - a_{21}\, r_2 - a_{31}\, r_3 - a_{41}\, r_4 + a_{13}\, r_1 + a_{23}\, r_2 + a_{33}\, r_3 + a_{43}\, r_4)$
$A_5 = 2\, (-\, a_{13}^2 - a_{23}^2 - a_{33}^2 - a_{43}^2 + a_{12}\, a_{13} + a_{22}\, a_{23} + a_{32}\, a_{33} + a_{42}\, a_{43} - a_{12}\, r_1 - a_{22}\, r_2 - a_{32}\, r_3 - a_{42}\, r_4 + a_{13}\, r_1 + a_{23}\, r_2 + a_{33}\, r_3 + a_{43}\, r_4)$
$A_6 = a_{13}^2 + a_{23}^2 + a_{33}^2 + a_{43}^2 + r_1^2 + r2^2 + r3^2 + r4^2 - 2 \times (a_{13}\, r_1 + a_{23}\, r_2 + a_{33}\, r_3 + a_{43}\, r_4)$

The values of the coefficients A_1 to A_6 are shown in Table 4.2.

The function to be minimized is:

$$F = A_1 x_1^2 + A_2 x_2^2 + A_3 x_1 x_2 + A_4 x_1 + A_5 x_2 + A_6 \tag{4.12}$$

In which the coefficients A_1 to A_6 are functions of the spectral values, a_{ij} (response values of the endmembers) and r_i (response values of the pixel).

To solve this problem, it is necessary to find a minimum value within the area defined by lines: $0 \leq x_1 \leq a$, $b \leq 0 \leq x_2$, and $x_1/a + x_2/b = 1$, where $a = b = 1$ (Fig. 4.2). Considering the function to be minimized, in order to find the minimum value, the partial derivatives are calculated and equated to zero:

$$\frac{dF}{dx_1} = 2A_1 x_1 + A_2 x_2 + A_4 = 0$$

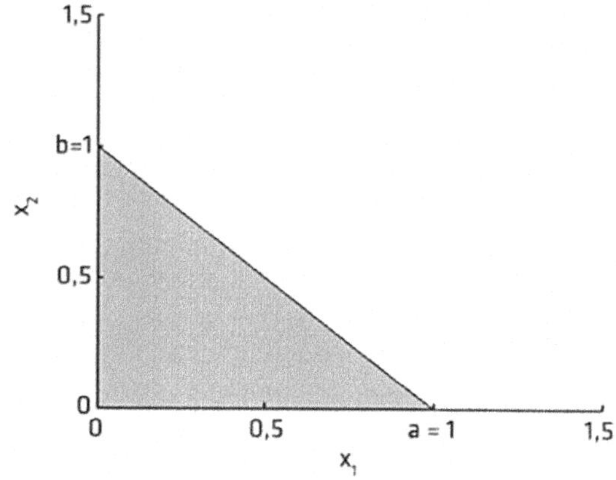

Fig. 4.2 Region that meets the restrictions for the number of components equal to 3

Table 4.3 Possible outcomes to the equation system solution

Outcome	X_1	X_2	Inside the region	Values to be recalculated	X_3
1	Positive	Positive	Yes		$1 - x_1 - x_2$
2	Positive	Positive	No	x_1 e x_2	0
3	Negative	Positive	No	x_2 ($x_1 = 0$)	$1 - x_2$
3	Negative	Negative	No	$x_1 = x_2 = 0$	1
3	Positive	Negative	No	x_1 ($x_2 = 0$)	$1 - x_1$

$$\frac{dF}{dx_2} = 2A_2x_2 + A_3x_1 + A_5 = 0$$

Solving for x_1 and x_2:

$$x_1 = \left(A_3A_5 - 2A_2A_4\right) / \left(4A_1A_2 - A_3^2\right)$$

$$x_2 = \left(A_3A_4 - 2A_1A_5\right) / \left(4A_1A_2 - A_3^2\right)$$

So, there are five possible outcomes (Table 4.3), which are described below.

Outcome 1 The minimum value is within the region of interest. So, this is the final solution and $x_3 = 1 - x_1 - x_2$.

Outcome 2 The minimum value is outside the region, and x_1 and x_2 are positive. In this case, the minimum value wanted is searched on the line defined by $x_1 + x_2 = 1$ (i.e., $x_3 = 0$). Now, making $x_2 = 1 - x_1$, the function to be minimized is:

$$F = \left(A_1 + A_2 - A_3\right)x_1^2 + \left(A_3 + A_4 - A_5 - 2A_2\right)x_1 + \left(A_2 + A_5 + A_6\right) \qquad (4.13)$$

The minimum value will be obtained by:

$$\frac{dF}{dx_1} = 2(A_1 + A_2 - A_3)x_1 + (A_3 + A_4 - A_5 - 2A_2) = 0$$

So:

$$x_1 = -\frac{(A_3 + A_4 - A_5 - 2A_2)}{2(A_1 + A_2 - A_3)}$$

If $x_1 > 1$, make $x_1 = 1 > 1$, or, if $x_1 < 0$, make $x_1 = 0$ and $x_2 = 1 - x_1$.

Outcome 3 The minimum value is outside the region and x_1 is negative and x_2 is positive. In this case, making $x_1 = 0$, the function to be minimized becomes:

$$F = A_2 x_2^2 + A_5 x_2 + A_6 \tag{4.14}$$

Solving to find the minimum, $x_2 = -A_5/2A_2$. If $x_2 > 1$, then make $x_2 = 1$, or, if $x_2 < 0$, make $x_2 = 0$ and $x_3 = 1 - x_2$.

Outcome 4 The minimum value is outside the region and x_1 and x_2 are negative. In this case, x_1 and x_2 are equated to zero and $x_3 = 1$.

Outcome 5 The minimum value is outside the region and x_1 is positive and x_2 is negative. In this case, making $x_2 = 0$, the function to be minimized becomes:

$$F = A_1 x_1^2 + A_4 x_1 + A_6 \tag{4.15}$$

Solving to find the minimum, $x_1 = -A_4/2A_1$. If $x_1 > 1$, then $x_1 = 1$, or, if $x_1 < 0$, then $x_1 = 0$ and $x_3 = 1 - x_1$.

4.1.3 Six Spectral Bands and Four Components

In this case, the mixture problem can be written as:

$$r_1 = a_{11}x_1 + a_{12}x_3 + a_{13}x_3 + a_{14}x_4 + e_1$$

$$r_2 = a_{21}x_1 + a_{22}x_3 + a_{23}x_3 + a_{24}x_4 + e_2$$

$$r_3 = a_{31}x_1 + a_{32}x_3 + a_{33}x_3 + a_{34}x_4 + e_3$$

$$r_4 = a_{41}x_1 + a_{42}x_3 + a_{43}x_3 + a_{44}x_4 + e_4$$

$$r_5 = a_{51}x_1 + a_{52}x_3 + a_{53}x_3 + a_{54}x_4 + e_5$$

$$r_6 = a_{61}x_1 + a_{62}x_3 + a_{63}x_3 + a_{64}x_4 + e_6$$

The function to be minimized is:

$$\begin{aligned} e_1^2 + e_2^2 + e_3^2 + e_4^2 + e_5^2 + e_6^2 = E_1x_1^2 + E_2x_2^2 + E_3x_3^2 + E_4x_4^2 + E_5x_1x_2 \\ + E_6x_1x_3 + E_7x_1x_4 + E_8x_2x_3 + E_9x_2x_4 \\ + E_{10}x_3x_4 + E_{11}x_1 + E_{12}x_2 \\ + E_{13}x_3 + E_{14}x_4 + E_{15} \end{aligned} \tag{4.16}$$

The values of the coefficients E_1 to E_{15} are shown in Table 4.4.

Consider the first restriction: $x_1 + x_2 + x_3 + x_4 = 1$ or $x_4 = 1 - x_1\ x_2\ x_3$. Replacing this restriction in Eq. 4.5, the function to be minimized becomes:

$$\begin{aligned} e_1^2 + e_2^2 + e_3^2 + e_4^2 + e_5^2 + e_6^2 = T_1x_1^2 + T_2x_2^2 + T_3x_3^2 + T_4x_1x_2 + T_5x_1x_3 \\ + T_6x_2x_3 + T_7x_1 + T_8x_2 + T_9x_3 + T_{10} \end{aligned} \tag{4.17}$$

in which the coefficients values of T_1 to T_{10} are shown in Table 4.5.

Now, the approach to solve this problem is to find a minimum within the volume defined by the plans: (a) $0 \leq x_1 \leq a$, $0 \leq x_2 \leq b$, and $x_1/a + x_2/b = 1$; (b) $0 \leq x_1 \leq a$, $0 \leq x_3$, and $x_1/a + x_3/c = 1$; (c) $0 \leq x_2 \leq b$, $0 \leq x_3 \leq c$, and $x_2/b + x_3/c = 1$; and (d) $x_1/a + x_2/b + x_3/c = 1$ (Spiegel 1968), in which $a = b = c = 1$ (Fig. 4.3).

Table 4.4 Coefficients values E_1 to E_{15} for Eq. 4.16

$E_1 = a_{11}^2 + a_{21}^2 + a_{31}^2 + a_{41}^2 + a_{51}^2 + a_{61}^2$
$E_2 = a_{12}^2 + a_{22}^2 + a_{32}^2 + a_{42}^2 + a_{52}^2 + a_{62}^2$
$E_3 = a_{13}^2 + a_{23}^2 + a_{33}^2 + a_{43}^2 + a_{53}^2 + a_{63}^2$
$E_4 = a_{14}^2 + a_{24}^2 + a_{34}^2 + a_{44}^2 + a_{54}^2 + a_{64}^2$
$E_5 = 2\ (a_{11}\ a_{12} + a_{21}\ a_{22} + a_{31}\ a_{32} + a_{41}\ a_{42} + a_{51}\ a_{52} + a_{61}\ a_{62})$
$E_6 = 2\ (a_{11}\ a_{13} + a_{21}\ a_{23} + a_{31}\ a_{33} + a_{41}\ a_{43} + a_{51}\ a_{53} + a_{61}\ a_{63})$
$E_7 = 2\ (a_{11}\ a_{14} + a_{21}\ a_{24} + a_{31}\ a_{34} + a_{41}\ a_{44} + a_{51}\ a_{54} + a_{61}\ a_{64})$
$E_8 = 2\ (a_{12}\ a_{13} + a_{22}\ a_{23} + a_{32}\ a_{33} + a_{42}\ a_{43} + a_{52}\ a_{53} + a_{62}\ a_{63})$
$E_9 = 2\ (a_{12}\ a_{14} + a_{22}\ a_{24} + a_{32}\ a_{34} + a_{42}\ a_{44} + a_{52}\ a_{54} + a_{62}\ a_{64})$
$E_{10} = 2\ (a_{13}\ a_{14} + a_{23}\ a_{24} + a_{33}\ a_{34} + a_{43}\ a_{44} + a_{53}\ a_{54} + a_{63}\ a_{64})$
$E_{11} = -2\ (a_{11}\ r_1 + a_{21}\ r_2 + a_{31}\ r_3 + a_{41}\ r_4 + a_{51}\ r_5 + a_{61}\ r_6)$
$E_{12} = -2\ (a_{12}\ r_1 + a_{22}\ r_2 + a_{32}\ r_3 + a_{42}\ r_4 + a_{52}\ r_5 + a_{62}\ r_6)$
$E_{13} = -2\ (a_{13}\ r_1 + a_{23}\ r_2 + a_{33}\ r_3 + a_{43}\ r_4 + a_{53}\ r_5 + a_{63}\ r_6)$
$E_{14} = -2\ (a_{14}\ r_1 + a_{24}\ r_2 + a_{34}\ r_3 + a_{44}\ r_4 + a_{54}\ r_5 + a_{64}\ r_6)$
$E_{15} = r_1^2 + r_2^2 + r_3^2 + r_4^2 + r_5^2 + r_6^2$

Table 4.5 Coefficients values of T_1 to T_{10} for Eq. 4.17

$T_1 = a_{11}^2 + a_{21}^2 + a_{31}^2 + a_{41}^2 + a_{51}^2 + a_{61}^2 + a_{14}^2 + a_{24}^2 + a_{34}^2 + a_{44}^2 + a_{54}^2 + a_{64}^2 - 2\,(a_{11}\,a_{14} + a_{21}\,a_{24} + a_{31}\,a_{34} + a_{41}\,a_{44} + a_{51}\,a_{54} + a_{61}\,a_{64})$
$T_2 = a_{12}^2 + a_{22}^2 + a_{32}^2 + a_{42}^2 + a_{52}^2 + a_{62}^2 + a_{14}^2 + a_{24}^2 + a_{34}^2 + a_{44}^2 + a_{54}^2 + a_{64}^2 - 2\,(a_{12}\,a_{14} + a_{22}\,a_{24} + a_{32}\,a_{34} + a_{42}\,a_{44} + a_{52}\,a_{54} + a_{62}\,a_{64})$
$T_3 = a_{13}^2 + a_{23}^2 + a_{33}^2 + a_{43}^2 + a_{53}^2 + a_{63}^2 + a_{14}^2 + a_{24}^2 + a_{34}^2 + a_{44}^2 + a_{54}^2 + a_{64}^2 - 2\,(a_{13}\,a_{14} + a_{23}\,a_{24} + a_{33}\,a_{34} + a_{43}\,a_{44} + a_{53}\,a_{54} + a_{63}\,a_{64})$
$T_4 = 2\,[(a_{11}\,a_{12} + a_{21}\,a_{22} + a_{31}\,a_{32} + a_{41}\,a_{42} + a_{51}\,a_{52} + a_{61}\,a_{62}) + (a_{14}^2 + a_{24}^2 + a_{34}^2 + a_{44}^2 + a_{54}^2 + a_{64}^2) - (a_{11}\,a_{14} + a_{21}\,a_{24} + a_{31}\,a_{34} + a_{41}\,a_{44} + a_{51}\,a_{54} + a_{61}\,a_{64}) - (a_{12}\,a_{14} + a_{22}\,a_{24} + a_{32}\,a_{34} + a_{42}\,a_{44} + a_{52}\,a_{54} + a_{62}\,a_{64})]$
$T_5 = 2\,[(a_{11}\,a_{13} + a_{21}\,a_{23} + a_{31}\,a_{33} + a_{41}\,a_{43} + a_{51}\,a_{53} + a_{61}\,a_{63}) + (a_{14}^2 + a_{24}^2 + a_{34}^2 + a_{44}^2 + a_{54}^2 + a_{64}^2) - (a_{11}\,a_{14} + a_{21}\,a_{24} + a_{31}\,a_{34} + a_{41}\,a_{44} + a_{51}\,a_{54} + a_{61}\,a_{64}) - (a_{13}\,a_{14} + a_{23}\,a_{24} + a_{33}\,a_{34} + a_{43}\,a_{44} + a_{53}\,a_{54} + a_{63}\,a_{64})]$
$T_6 = 2\,[(a_{12}\,a_{13} + a_{22}\,a_{23} + a_{32}\,a_{33} + a_{42}\,a_{43} + a_{52}\,a_{53} + a_{62}\,a_{63}) + (a_{14}^2 + a_{24}^2 + a_{34}^2 + a_{44}^2 + a_{54}^2 + a_{64}^2) - (a_{12}\,a_{14} + a_{22}\,a_{24} + a_{32}\,a_{34} + a_{42}\,a_{44} + a_{52}\,a_{54} + a_{62}\,a_{64}) - (a_{13}\,a_{14} + a_{23}\,a_{24} + a_{33}\,a_{34} + a_{43}\,a_{44} + a_{53}\,a_{54} + a_{63}\,a_{64})]$
$T_7 = -2\,[(a_{11}\,r_1 + a_{21}\,r_2 + a_{31}\,r_3 + a_{41}\,r_4 + a_{51}\,r_5 + a_{61}\,r_6) + (a_{14}^2 + a_{24}^2 + a_{34}^2 + a_{44}^2 + a_{54}^2 + a_{64}^2) - (a_{11}\,a_{14} + a_{21}\,a_{24} + a_{31}\,a_{34} + a_{41}\,a_{44} + a_{51}\,a_{54} + a_{61}\,a_{64}) - (a_{14}\,r_1 + a_{24}\,r_2 + a_{34}\,r_3 + a_{44}\,r_4 + a_{54}\,r_5 + a_{64}\,r_6)]$
$T_8 = -2\,[(a_{12}\,r_1 + a_{22}\,r_2 + a_{32}\,r_3 + a_{42}\,r_4 + a_{52}\,r_5 + a_{62}\,r_6) + (a_{14}^2 + a_{24}^2 + a_{34}^2 + a_{44}^2 + a_{54}^2 + a_{64}^2) - (a_{12}\,a_{14} + a_{22}\,a_{24} + a_{32}\,a_{34} + a_{42}\,a_{44} + a_{52}\,a_{54} + a_{62}\,a_{64}) - (a_{14}\,r_1 + a_{24}\,r_2 + a_{34}\,r_3 + a_{44}\,r_4 + a_{54}\,r_5 + a_{64}\,r_6)]$
$T_9 = -2\,[(a_{13}\,r_1 + a_{23}\,r_2 + a_{33}\,r_3 + a_{43}\,r_4 + a_{53}\,r_5 + a_{63}\,r_6) + (a_{14}^2 + a_{24}^2 + a_{34}^2 + a_{44}^2 + a_{54}^2 + a_{64}^2) - (a_{13}\,a_{14} + a_{23}\,a_{24} + a_{33}\,a_{34} + a_{43}\,a_{44} + a_{53}\,a_{54} + a_{63}\,a_{64}) - (a_{14}\,r_1 + a_{24}\,r_2 + a_{34}\,r_3 + a_{44}\,r_4 + a_{54}\,r_5 + a_{64}\,r_6)]$
$T_{10} = r_1^2 + r_2^2 + r_3^2 + r_4^2 + r_5^2 + r_6^2 + a_{14}^2 + a_{24}^2 + a_{34}^2 + a_{44}^2 + a_{54}^2 + a_{64}^2 - 2\,(a_{14}\,r_1 + a_{24}\,r_2 + a_{34}\,r_3 + a_{44}\,r_4 + a_{54}\,r_5 + a_{64}\,r_6)$

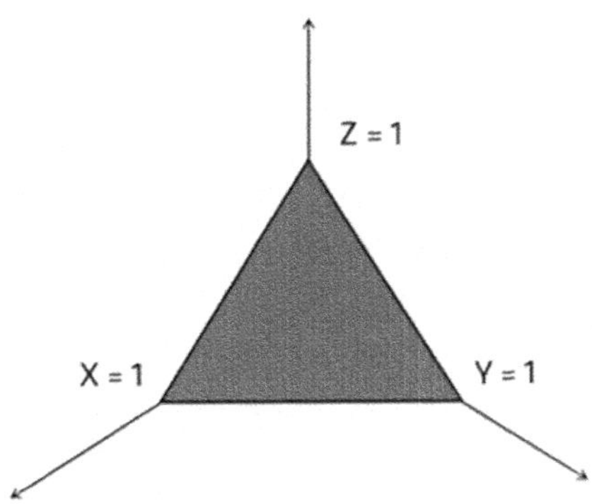

Fig. 4.3 Region that meets the restrictions for the number of components equal to 4

The function to be minimized is:

$$F = T_1x_1^2 + T_2x_2^2 + T_3x_3^2 + T_4x_1x_2 + T_5x_1x_3 + T_6x_2x_3 + T_7x_1 + T_8x_2 + T_9x_3 + T_{10} \quad (4.18)$$

To find the minimum, the partial derivatives are calculated and equated to zero:

$$\frac{dF}{dx_1} = 2T_1x_1 + T_4x_2 + T_5x_3 + T_7 = 0$$

$$\frac{dF}{dx_2} = 2T_2x_2 + T_4x_1 + T_6x_3 + T_8 = 0$$

$$\frac{dF}{dx_3} = 2T_3x_3 + T_5x_1 + T_6x_2 + T_9 = 0$$

Solving for x_1, x_2, and x_3, the system of linear equations can be formulated as:

$$\begin{vmatrix} 2T_1 & T_4 & T_5 \\ T_4 & 2T_2 & T_6 \\ T_5 & T_6 & 2T_3 \end{vmatrix} \begin{vmatrix} x_1 \\ x_2 \\ x_3 \end{vmatrix} = \begin{vmatrix} -T_7 \\ -T_8 \\ -T_9 \end{vmatrix}$$

Using a numerical method to solve the system of linear equations (Burden et al. 1981; Conte and De Boor 1980), the minimum not restricted can be obtained. Thus, there are nine possible outcomes (Table 4.6).

The procedure to calculate the values of x_1, x_2, x_3, and x_4 is therefore as follows:

Outcome 1 $0 \leq x_1 \leq 1, 0 \leq x_2 \leq 1, 0 \leq x_3 \leq 1$, and $x_1 + x_2 + x_3 \leq 1$, that is, the minimum is within the region of interest. So, this is the final solution and $x_4 = 1 - (x_1 + x_2 + x_3)$.

Table 4.6 Possible outcomes or the solution of system of equations

Outcome	x_1	x_2	x_3	Inside the region	Values to be recalculated	x_4
1	Positive	Positive	Positive	Yes	–	$1 - x_1 - x_2 - x_3$
2	Positive	Positive	Positive	No	x_1, x_2, x_3	0
3	Negative	Positive	Positive	No	x_2, x_3 ($x_1 = 0$)	$1 - x_2 - x_3$
4	Positive	Negative	Positive	No	x_1, x_3 ($x_2 = 0$)	$1 - x_1 - x_3$
5	Negative	Negative	Positive	No	x_3 ($x_1 = x_2 = 0$)	$1 - x_3$
6	Positive	Positive	Negative	No	x_1, x_2 ($x_3 = 0$)	$1 - x_1 - x_2$
7	Negative	Positive	Negative	No	x_2 ($x_1 = x_3 = 0$)	$1 - x_2$
8	Negative	Negative	Negative	No	($x_1 = x_2 = x_3 = 0$)	1
9	Positive	Negative	Negative	No	x_1 ($x_2 = x_3 = 0$)	$1 - x_1$

Outcome 2 The minimum is outside the region of interest, and x_1, x_2, and x_3 are positive. In this case, the problem is to find the minimum in the plane defined by $x_1 + x_2 + x_3 = 1$, that is, $x_4 = 0$. So, x_3 can be replaced by $(1 - x_1 - x_2)$ and the function to be minimized becomes:

$$F = U_1 x_1^2 + U_2 x_2^2 + U_3 x_1 x_2 + U_4 x_1 + U_5 x_2 + U_6 \tag{4.19}$$

where:

$$U_1 = T_1 + T_3 - T_5$$

$$U_2 = T_2 + T_3 - T_6$$

$$U_3 = 2T_3 + T_4 - T_5 - T_6$$

$$U_4 = T_5 + T_7 - T_9 - 2T_3$$

$$U_5 = T_6 + T_8 - T_9 - 2T_3$$

$$U_6 = T_3 + T_9 + T_{10}$$

The function to be minimized is similar to the case of three components presented earlier. Then, x_1 and x_2 are calculated according to the procedure described and $x_3 = 1 - x_1 - x_2$ and $x_4 = 0$.

Outcome 3 The minimum is outside the region of interest, x_2 and x_3 are positive, and x_1 is negative. In this case, making $x_1 = 0$, the function to be minimized becomes:

$$F = U_1 x_2^2 + U_2 x_3^2 + U_3 x_2 x_3 + U_4 x_2 + U_5 x_3 + U_6 \tag{4.20}$$

where:

$$U_1 = T_2$$

$$U_2 = T_3$$

$$U_3 = T_6$$

$$U_4 = T_8$$

$$U_5 = T_9$$

$$U_6 = T_{10}$$

The function to be minimized is similar to the case of three components presented earlier. So, x_2 and x_3 are calculated according to the procedure described and $x_4 = 1 - x_2 - x_3$ and $x_1 = 0$.

Outcome 4 The minimum is outside the region of interest, x_1 and x_3 are positive, and x_2 is negative. In this case, making $x_2 = 0$, the function to be minimized becomes:

$$F = U_1x_1^2 + U_2x_3^2 + U_3x_1x_3 + U_4x_1 + U_5x_3 + U_6$$

where:

$$U_1 = T_1$$

$$U_2 = T_3$$

$$U_3 = T_5$$

$$U_4 = T_7$$

$$U_5 = T_9$$

$$U_6 = T_{10}$$

The function to be minimized is similar to the case of three components presented earlier. Then, x_1 and x_3 are calculated according to the procedure described and $x_4 = 1 - x_1 - x_3$ and $x_2 = 0$.

Outcome 5 The minimum is outside the region of interest, x_1 and x_2 are negative, and x_3 is positive. In this case, making $x_1 = x_2 = 0$, the function to be minimized becomes:

$$F = T_3x_3^2 + T_9x_3 + T_{10} \quad (4.22)$$

And, to find the minimum:

$$\frac{dF}{dx_3} = 2T_3x_3 + T_9 = 0$$

Then:

$$x_3 = -T_9 / 2T_3$$

If x_3 is in the range between 0 and 1, so this is the final solution. If x_3 is greater than 1, make $x_3 = 1$, or, if x_3 is less than 0, make $x_3 = 0$ and $x_4 = 1 - x_3$.

Outcome 6 The minimum is outside the region of interest, x_1 and x_2 are positive, and x_3 is negative. In this case, making $x_3 = 0$, the function to be minimized becomes:

$$F = U_1x_1^2 + U_2x_2^2 + U_3x_1x_2 + U_4x_1 + U_5x_2 + U_6 \quad (4.23)$$

where:

$$U_1 = T_1$$

$$U_2 = T_2$$

$$U_3 = T_4$$

$$U_4 = T_7$$

$$U_5 = T_8$$

$$U_6 = T_{10}$$

The function to be minimized is similar to the case of three components presented earlier. Then, x_1 and x_2 are calculated according to the procedure described and $x_4 = 1 - x_1 - x_2$ and $x_3 = 0$.

Outcome 7 The minimum is outside the region of interest, x_2 is positive, and $x1$ and x_3 are negative. In this case, making $x_1 = x_3 = 0$, the function to be minimized becomes:

$$F = T_2 x_2^2 + T_8 x_2 + T_{10} \tag{4.24}$$

And, to find the minimum:

$$\frac{dF}{dx_2} = 2T_2 x_2 + T_8 = 0$$

Then:

$$x_1 = -T_7 / 2T_1$$

If x_2 is in the range between 0 and 1, then this is the final solution. If x_2 is greater than 1, make $x_2 = 1$, or, if x_2 is less than 0, make $x_2 = 0$ and $x_4 = 1 - x_2$.

Outcome 8 The minimum is outside the region of interest, and x_1, x_2, and x_3 are negative. In this case, making $x_1 = x_2 = x_3 = 0$, then $x_4 = 1$.

Outcome 9 The minimum is outside the region of interest, x_1 is positive, and x_2 and x_3 are negative. In this case, making $x_2 = x_3 = 0$, the function to be minimized becomes:

$$F = T_1 x_1^2 + T_7 x_1 + T_{10} \tag{4.25}$$

And, to find the minimum:

$$\frac{dF}{dx_1} = 2T_1x_1 + T_7 = 0$$

Then:

$$x_1 = -T_7 / 2T_1$$

If x_1 is in the range between 0 and 1, so this is the final solution. If x_1 is greater than 1, make $x_1 = 1$, or, if x_1 is less than 0, make $x_1 = 0$ and $x_4 = 1 - x_1$.

4.1.4 Weighted Least Squares (WLS)

Consider the curve fit the data with a curve having the form:

$$R = f\left(A,,,,x_1,,,,x_2,,,,\ldots,,,,x_n\right) = x_1 f\left(A\right) + x_2 f\left(A\right) + \ldots + x_n f\left(A\right) \quad (4.26)$$

where the dependent variable R is linear with respect to x_1, x_2, ..., x_n.

Although there are many ramifications and approaches to the curves fitting, the least squares method can be applied to a wide variety of curves fitting problems involving linear form with undetermined values. The values are determined by minimizing the sum of squared errors (residuals). The obtained solution by this method is mathematically possible, but an example of what is mentioned being physically unacceptable (some restrictions are involved: the values should not be negative and should add to 1). Then, it becomes a problem of constrained least squares, and the equations of constraints should be added. To solve this problem, it is necessary to apply the concepts of weighted least squares.

Sometimes, the information obtained in an experiment can be more accurate than those arising from other sources of information related to the same experiment. In other cases, it is convenient to use some additional information (previous knowledge) to make the solution physically relevant. In such cases, it may be desirable to give a higher "weight" for the information that are considered more accurate or more important to the problem. To weigh certain information (e.g., additional information) is desirable to bring the solution near the physical meaning, then getting an acceptable solution.

In this case, $x_1 + x_2 + \ldots + x_n = 1$ and $0 \leq x_1, x_2, \ldots, x_n \leq 1$ are the conditions that must be met to obtain an acceptable solution. So, $n + 1$ equations are added to the system of Eq. 4.8: one corresponding to the conditional sum of proportions equal to 1 ($x_1 + x_2 + \ldots + x_n = 1$) and other n corresponding to the condition that the proportions should not be negative ($x_j \leq 1, j = 1, 2, \ldots, n$). To resolve this issue, when restrictions are not met, it is applied a diagonal matrix W containing values of weights associated with the system of equations to be solved. Initially the first assigned values equal to 1, along the diagonal matrix W, mean that the equations are equally important for the solution of the problem. The very high value assigned to the diagonal corresponding to the first constraint (sum of $x_j = 1$) indicates that this equation must be strictly satisfied. So, if the values of x_j's are satisfied, that is, if they are in the range between 0 and 1, then the final solution was found. Otherwise, an iterative process must be used in order to bring all x_j's within the range between 0 and 1. This is accomplished by a gradual increase of the weights, which are initially zero, corresponding to n latest equations relating to the restriction that the proportions should not be negative. The solution to this problem is found by minimizing the amount: $W_1 e_1^2 + W_2 e_2^2 + \ldots + W_{(m+n+1)} e_{(m+n+1)}^2$, where W_1, W_2, etc. are the weight factors and e_1, e_2, etc. are the error values for each equation.

The implementation of this method is based on Gaussian elimination and replacement algorithm (forward and backward), described in textbooks of numerical analysis, such as Burden et al. (1981).

4.1.5 Principal Components

Given an image consisting of a number of pixels with measures in a number of spectral bands, it is possible to model each spectral response of each pixel as a linear combination of a finite number of components.

$$
\begin{array}{ccccccc}
dn_1 & = & f_1 e_{1,1} + & \cdots & + f_n e_{1,n} & \text{band } 1 \\
dn_2 & = & f_1 e_{2,1} + & \cdots & + f_n e_{2,n} & \text{band } 2 \\
\vdots & & \vdots & & \vdots & \vdots \\
dn_p & = & f_1 e_{p,1} + & \cdots & + f_n e_{p,n} & \text{band } p
\end{array}
$$

where:

d_{nj} = DN for the band i of the pixel
$e_{i,j}$ = pure component dn of the pure component j, band i
f_j = unknown fraction of the pure component j
n = number of pure components
p = number of bands

That makes the matrix equation:

$$dn = ef \tag{4.27}$$

A linear constraint is added because the sum of the fractions of any pixel must be equal to 1; therefore, it needs to increase the vector dn with an additional 1 and the matrix *e* with a row of values 1. This becomes a set of $p + 1$ equations in n unknowns. Since the number of pure components is generally smaller than the number of spectral bands, the equations are possibly undetermined and can be solved by any other techniques. The solution described uses principal component analysis (PCA) to reduce the dimensionality of the dataset. The matrix of pure component is transformed into a PCA space using the appropriate number of eigenvectors, the pixel data are transformed into PCA space, the solutions are found, and the resulting fractions are stored.

The weighted least squares method and the principal component method are recommended for the cases where the number of spectral components in the mixture is greater than three.

4.2 Endmembers Selection

To generate the fraction images, it is necessary to choose the pure components (endmembers) to apply any available mathematical algorithm. Explicitly defined, the endmember is just a component that is part of the spectral mixture. So, it is necessary to choose the endmembers that make sense for the interpretation of the image considered and also that meet the criteria of the fraction according to the equations of the mixture already presented earlier. Sometimes, it is easy to choose these endmembers because targets are already known for the area being studied. This is true for areas that have been studied for some time, for example, when the changes in the environment are being monitored. On the other hand, it is necessary to perform experiments to find the appropriate endmembers whenever the scenes are unknown or when it is needed to extract specified information from the images.

There are two ways to select the endmembers: directly from the images and through collections of data obtained in the laboratory and/or in the field. The endmembers derived from the images are called image endmembers, while those selected from laboratory data and/or in the field are called reference endmembers. The most convenient is to select the endmembers directly from the images being studied for the simple reason that the spectrum of the endmember extracted from the image can be used without calibration. In this case, spectra obtained in the laboratory or in the field present values of reflectance factors. As seen earlier, the orbital images are, in principle, available in the DN format. In that way, the reference endmembers (reflectance factors) would be in scale or measure unit different from that adopted by the images (DNs), which would interfere in the application of the mixture model. The right procedure, then, would be to convert the DNs of images in surface reflectance factors, with atmospheric correction, in order to have compatibility on the unity of the two data sets.

Image endmembers, although convenient for not needing the calibration, do not always work in mixture models. To work well, it is necessary to have a good relationship

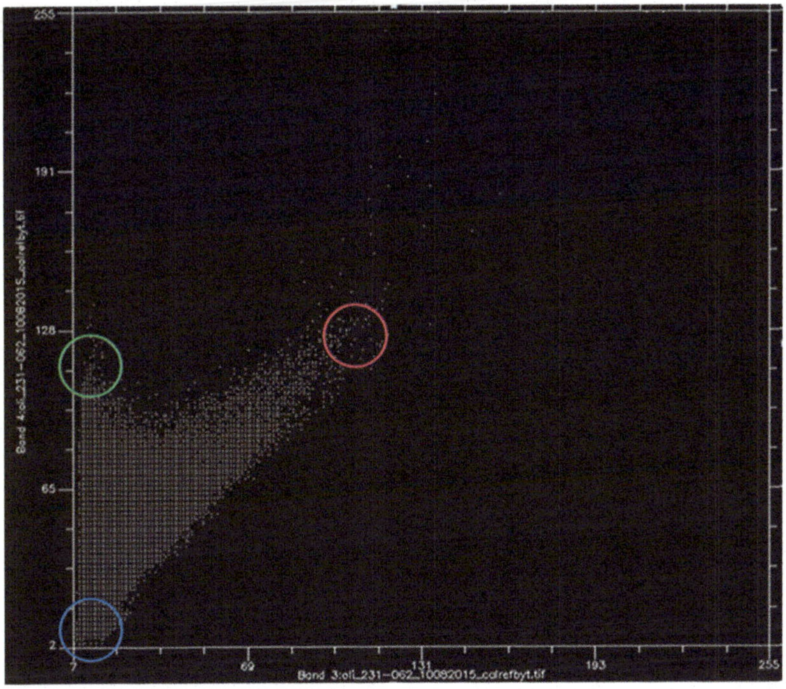

Fig. 4.4 Example of dispersion of pixels of an image in the graph formed by the red and near-infrared bands, showing the potential endmembers of vegetation (green), soil (red), and shadow/water (blue)

between the pixel scale and the scale in which the materials occur relatively pure on the ground. In the best case, it can be used image endmember if an image present at least some pixels fully occupied by a pure material on the ground.

It can be concluded in a way more or less intuitive that the definition of the number and the selection of endmembers to be considered are essential to the success of the application of mixture model. Nevertheless, it is known that, in the real world, the terrain can be spectrally complex. One reason why the images of scenes on the ground can be well modeled by few endmembers spectra is due to the fact that some of the potential endmembers is in small proportions in comparison with the endmembers considered in the mixture. In this way, the determination of the number of endmembers is made by defining the vertices of a geometric figure that encompass the spectral responses of the pixels of the scene. For example, in the case of three endmembers, the geometric figure is a triangle on the two-dimensional plane formed by two spectral bands (in general, for the land cover, red and near-infrared spectral bands of the electromagnetic spectrum) (Fig. 4.4). The endmembers will be those in which the spectral responses are closest to the vertices of the geometric figure formed.

Now, in the event that it is decided to use image endmembers, there will be fraction images with the proportion of 100% (pure pixel), while, if it is decided to use reference endmembers, it will be hardly to have any pure pixel in the fraction images.

Chapter 5
Fraction Images

Abstract This chapter is dedicated to the presentation of some linear spectral mixture model applications that is based in the so-called fraction images. Details are also presented on the interpretation of those images and their meanings.

Keywords Fraction images · Vegetation fraction · Soil fraction · Shade/water fraction

The fraction images are products generated from the application of the mathematical algorithms described previously. They represent the proportions of the components in the spectral mixture. In general, all algorithms produce the same result, i.e., generate the same fraction images when the constraint equations are not used or the proportions are in the range from 0 to 1. Normally the vegetation, soil, and shade/water fraction images are generated, which are the components usually present in any scene on the ground. The fraction images can be considered a form of dimensionality reduction of the data and also a form of enhancement of the information. In addition, the spectral mixture model transforms the spectral information into physical information (proportion values of the components in the pixel; not to be confused with another type of transformation that converts spectral or radiometric data into physical unities as radiance or reflectance).

The vegetation fraction image highlights the areas of vegetation cover, the soil fraction image highlights the areas of exposed soil, and the shade/water fraction image highlights the occupied areas with water bodies such as rivers and lakes and also the burned areas, the wetlands, etc. The shade and the water are considered together because these two targets exhibit similar responses in the spectral bands typically used by Earth observation sensors. In this way, it is important to keep on mind that the spectral mixture model is not a classifier, but rather an image transformation technique to facilitate the extraction of information.

For the generation of fraction images, the spectral responses of pure components (endmembers) are considered known, i.e., can be obtained directly from the images (image endmember) or from the available spectral libraries (reference endmember). Figure 5.1 shows an example of the spectral responses of vegetation, soil, and shade/

Y. E. Shimabukuro, F. J. Ponzoni, *Spectral Mixture for Remote Sensing*,
Springer Remote Sensing/Photogrammetry,
https://doi.org/10.1007/978-3-030-02017-0_5

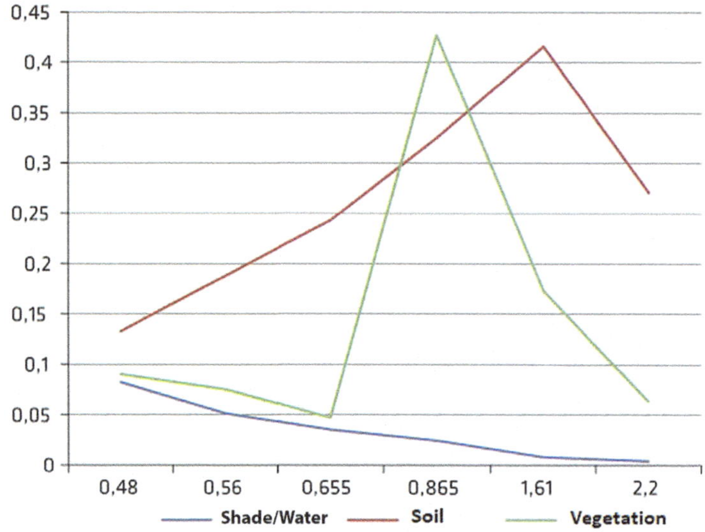

Fig. 5.1 Spectral response of vegetation (green color), soil (red color), and shade/water (blue color) components

water components used to generate fraction images in a scene of the OLI/Landsat 8 image Path 231/row 062, corresponding to the Manaus (AM) region. In this case it was used the images of all bands, 1 (0.43 μm–0.45 μm), 2 (0.45 μm–0.51 μm), 3 (0.59 μm–0.53 μm), 4 (0.64–0.67 μm), 5 (0.85 μm–0.88 μm), 6 (1.65 μm–1.57 μm), and 7 (2.29 μm–2.11 μm), that were previously converted to apparent reflectance values. It is worth to remember that this analysis could be performed using these values of apparent or surface reflectance or even DNs.

Obviously, the users must take into account the data type which they are working, particularly when selecting the endmembers. Users with more familiarity with the analysis of reflectance curves of different natural resources probably will feel more comfortable to select endmembers when working with images converted into surface reflectance values, since the shape of the curves will inform about the nature of the pixels selected as pure. It does not mean that they are not able to make good selection working with images composed by DNs. In this case the shape of the curves would not be very useful, but it would not interfere in the performance of the mixture model.

After applying the mixture model, new images are generated, composed then by numbers that represent the proportion values of a given component within each pixel. As an example, Fig. 5.2a shows a color composite (R6 G5 B4) based on images from the OLI/Landsat 8 sensor, while Fig. 5.2b–g presents the corresponding images of the bands 2–7.

Figure 5.3a shows a color composite of the fraction images Soil Fig. 5.3b, Vegetation Fig. 5.3c and Shade/water Fig. 5.3d derived from the sensor OLI/Landsat 8 for the Manaus region (AM) soil, vegetation, and shade/water fraction

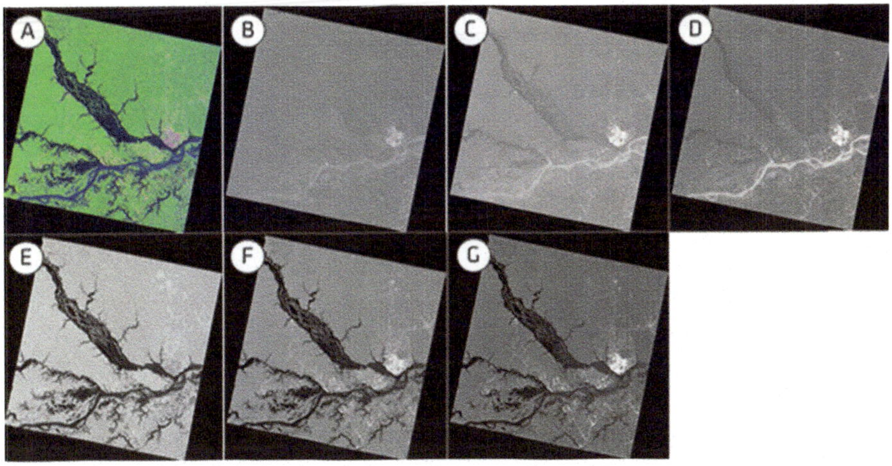

Fig. 5.2 (**a**) Color composite (R6 G5 B4) of OLI/Landsat 8 for 231/062 image; (**b**) band 2; (**c**) band 3; (**d**) band 4; (**e**) band 5; (**f**) band 6; (**g**) band 7

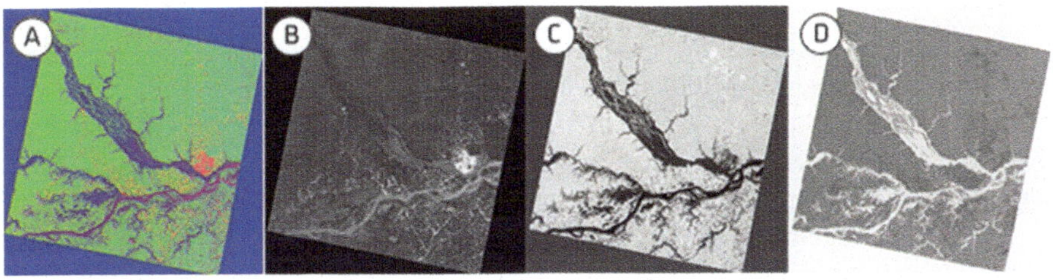

Fig. 5.3 (**a**) Color composite of the fraction images for the region of Manaus (AM) and the fraction images (**b**) soil, (**c**) vegetation, and (**d**) shade/water

In the vegetation fraction image (Fig. 5.3c), the brighter pixels are those which, at least in theory, have greater amount of vegetation, while the water bodies appear dark since they do not have any percentage of vegetation cover. A similar analysis can be done with the other components of fraction images. For example, in the soil fraction image (Fig. 5.3b), the brighter pixels are those that have the lowest levels of vegetation cover or are less shaded.

Figure 5.4 shows a color composite (R6 G2 B1) of the bands 1 (red), 2 (near infrared), and 6 (medium infrared) of MODIS/Terra sensor for the west region of the state of São Paulo in Brazil. The corresponding vegetation, soil, and shade/water fraction images are presented in Fig. 5.4b–d.

Figure 5.5 displays a color composite (R6 G5 B4) of the bands 4 (red), 5 (near infrared), and 6 (medium infrared) of the OLI Landsat 8 sensor for the part of the image Path 226/row 068 in the state of Mato Grosso, while Fig. 5.5b–d presents the corresponding vegetation, soil, and shade/water fraction images.

It is observed that the fraction images are monochromatic (grayscale) and the DNs are directly associated with the proportions (abundance) of each of the

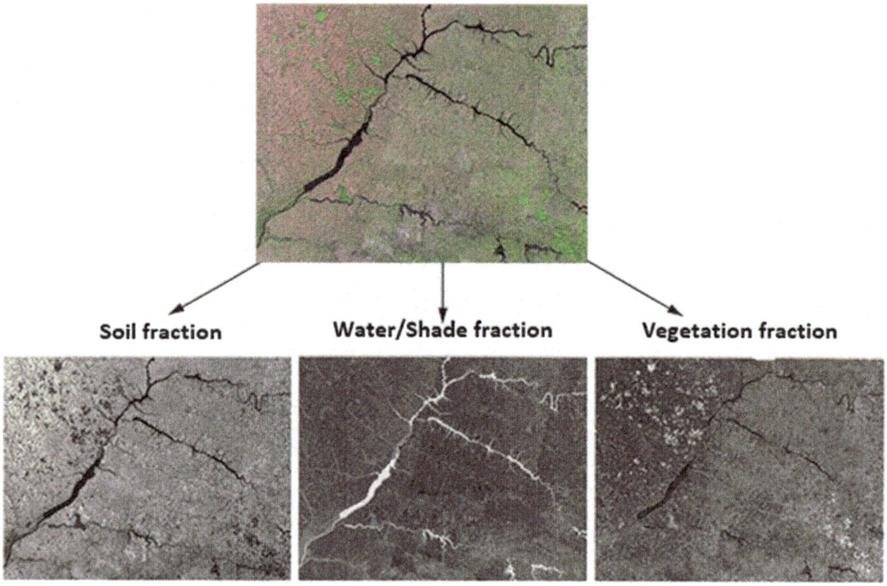

Fig. 5.4 (**a**) Color composite (R6 G2 B1) of MODIS/Terra for the west region of the state of São Paulo and the fraction images (**b**) soil, (**c**) shade/water, and (**d**) vegetation

respective components of the scene selected for the spectral mixture model. Thus, the higher the DN value in the vegetation fraction image (Fig. 5.5b), the greater the proportion of vegetation in the corresponding pixel (bright green in Fig. 5.5). The same interpretation applies to the other components fraction images: the higher the DN value in the soil fraction image (Fig. 5.5c), the greater the proportion of soil in the corresponding pixel (magenta in Fig. 5.5), and the higher the DN value in the shade/water fraction image (Fig. 5.5d), the greater the proportion of water or burned in the corresponding pixel (dark magenta or black in Fig. 5.5a).

The literature presents a large amount of works on the use of the linear spectral mixture model in several regions around the world, showing that this technique is consistent. In addition, the fraction images generated by this model are being used in different application areas, such as forest, agriculture, land use, water, and urban areas.

So, the proportion of each endmember can be shown for each pixel, creating an image useful for photointerpretation. Figures 5.3, 5.4, and 5.5 show the proportions of each endmember represented in gray levels. The fraction images are derived on the basis of the information from all multispectral bands used. For each type of application, a limited number of spectral bands can be employed, for example, the deforested area analysis in the Amazon region can be based on only three bands: red, near infrared, and short wave infrared.

Therefore, the conversion of the spectral data in fraction images by means of linear spectral mixture model can result in a significant reduction in dimensionality of the data to be analyzed. For example, it is possible to use several bands of different sensors – 6 bands of TM, 7 bands of MODIS, and 242 bands of Hyperion – to generate a small number of fraction images (usually, three or four endmembers).

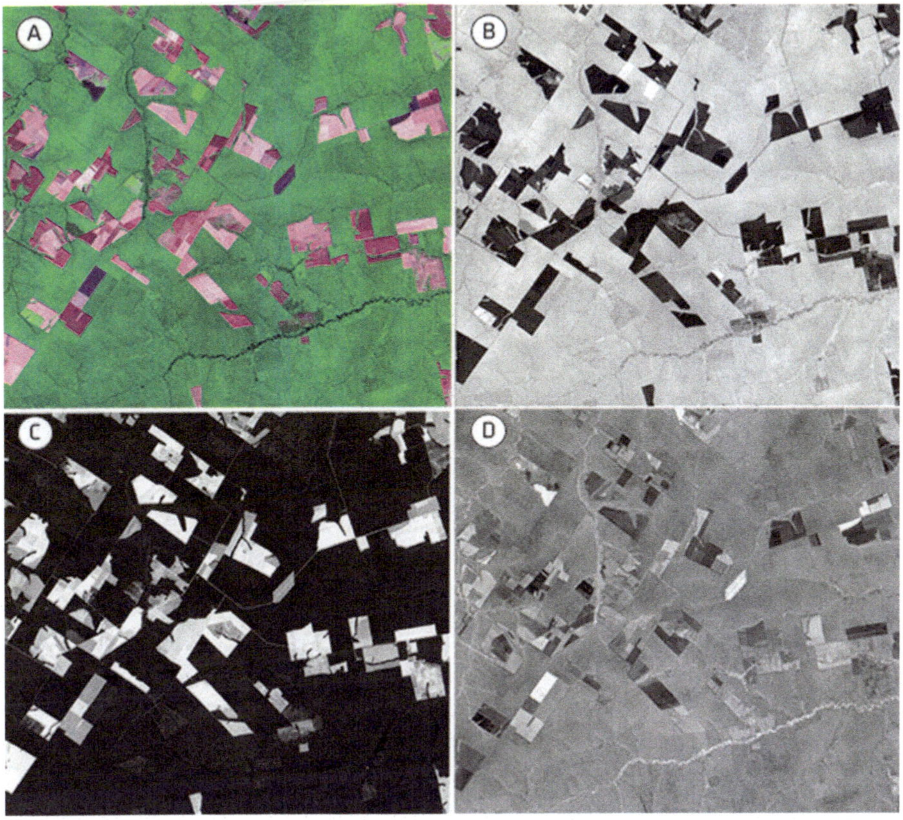

Fig. 5.5 (**a**) Color composite (R6 G5 B4) of OLI/Landsat 8 of the part of the image Path 226/row 068 in the state of Mato Grosso and fraction images (**b**) vegetation, (**c**) soil, and (**d**) shade/water

Now, once known the proportions of the fraction images and the spectral responses of the endmembers, it is possible to retrieve the spectral responses of the pixels in each of the spectral bands used. This procedure allows evaluating the performance of the models by generating the *error images*, which will be presented in the next section.

The grayscale of fraction images are brighter with the higher proportion of endmember in the pixel. The same convention is adopted for the error image associated with the model. The choice of the convention is arbitrary, so that sometimes, the tones of the image can be reversed for helping the visualization of the patterns, for example, in the case of the shade/water fraction image, it is usually more intuitive when the tones are reversed, so that dark tones indicate less lighting. The fraction images can also be presented in color composite (RGB) by selecting three images corresponding to the endmembers. In this case, the image contrast stretching can be applied for the visualization of the patterns but distorts the proportion values for the quantitative information.

To analyze a fraction image individually, it is more effective to present the higher proportions of the endmembers in brighter grayscales, to highlight the target of interest.

In case of considering three endmembers such as vegetation, soil, and shade/water, the vegetation image enhances the vegetation cover so proportional to the vegetation greenness, the soil fraction image highlights the areas without vegetation cover, and the shade/water fraction image highlights the areas occupied by water bodies and the burned areas.

5.1 Error Images

As mentioned earlier, based on the spectral mixture model, it is possible to calculate the error for each of the spectral bands and to generate the corresponding error images, since the responses of the components and their proportions are known. This is one way to evaluate the performance of the model, that is, when the model is appropriate, the error images show a nonstandard aspect. If there is a component that has not been considered in the mixture, it will be highlighted in the error images of the spectral bands employed.

It is possible to write the mixture model as described earlier:

$$\begin{array}{lllllll} r_1 = & a_{11}x_1 + & a_{12}x_2 + & \cdots & +a_{1n}x_n & +e_1 \\ r_2 = & a_{21}x_1 + & a_{22}x_2 + & \cdots & +a_{2n}x_n & +e_2 \\ \vdots & \vdots & \vdots & & \vdots & \vdots \\ r_m = & a_{m1}x_1 + & a_{m2}x_2 + & \cdots & +a_{mn}x_n & +e_m \end{array}$$

or,

$$r_i = \Sigma\left(a_{ij}x_j\right) + e_i \tag{5.1}$$

So the errors for each band can be obtained by:

$$\begin{array}{lllllll} e_1 = & r_1 - & (a_{11}x_1 + & a_{12}x_2 + & \cdots & +a_{1n}x_n) \\ e_2 = & r_2 - & (a_{21}x_1 + & a_{22}x_2 + & \cdots & +a_{2n}x_n) \\ \vdots & \vdots & \vdots & \vdots & & \vdots \\ e_m = & r_m - & (a_{m1}x_1 + & a_{m2}x_2 + & \cdots & +a_{mn}x_n) \end{array}$$

or,

$$e_i = r_i - \Sigma\left(a_{ij}x_j\right) \tag{5.2}$$

where:

r_i = average spectral reflectance for the i-th spectral band
a_{ij} = spectral reflectance of the j-th component in the pixel for the i-th spectral band

x_j = proportion value of the j-th component in the pixel
e_i = error for the i-th spectral band
j = 1, 2,..., n (n = number of components assumed for the problem)
i = 1, 2,..., m (m = number of spectral bands for the sensor system)

The error images usually present low values according to the accuracy of the employed models. In this way, these images are used to evaluate the quality of the defined models, that is, if the number of components is suitable for the analyzed scene. If there is any component not represented in the mixture, it will be highlighted in these error images.

Figures 5.6 and 5.7 present an example of the evaluation of the spectral mixture model using an OLI/Landsat 8 image obtained over an area of the state of Mato Grosso (path 226/row 068).

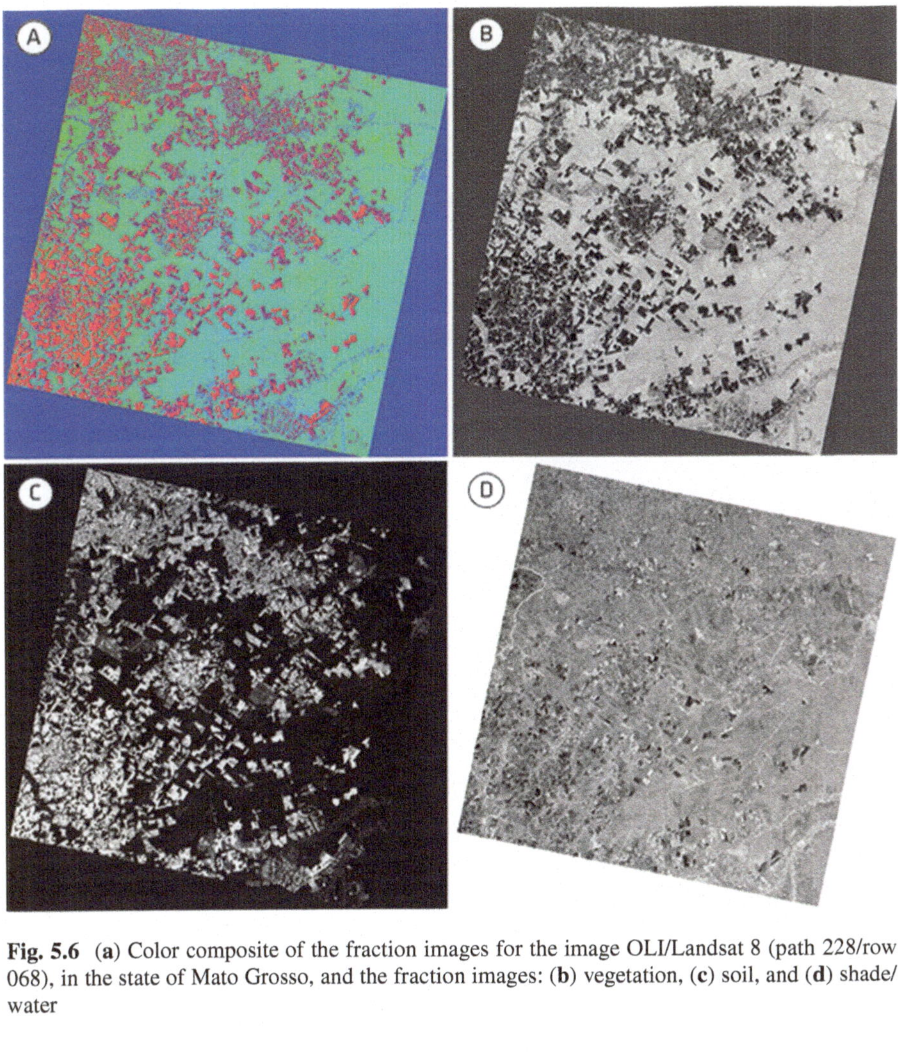

Fig. 5.6 (**a**) Color composite of the fraction images for the image OLI/Landsat 8 (path 228/row 068), in the state of Mato Grosso, and the fraction images: (**b**) vegetation, (**c**) soil, and (**d**) shade/ water

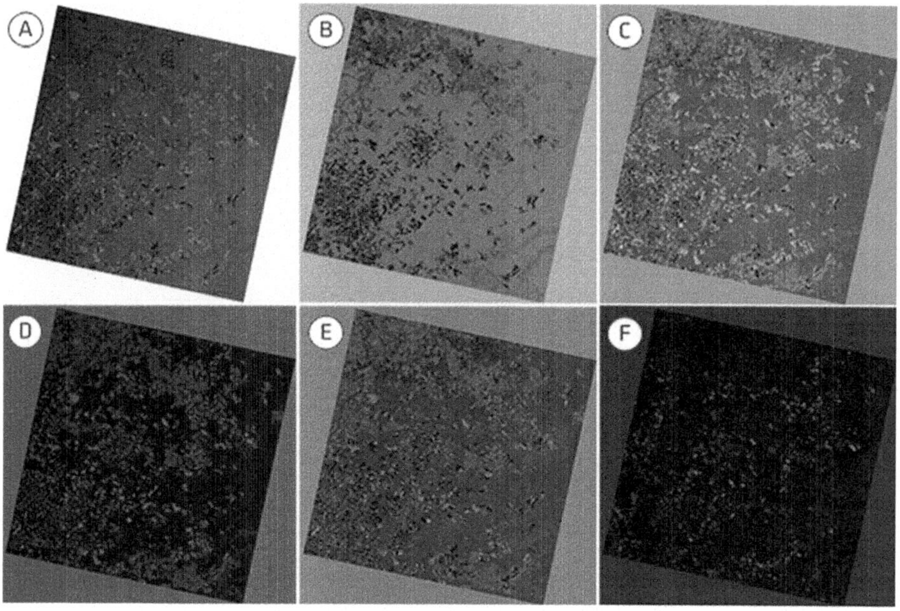

Fig. 5.7 Error images: (**a**) band 2; (**b**) band 3; (**c**) band 4; (**d**) band 5; (**e**) band 6; (**f**) band 7

In this example, considering all bands, the average error was 6.109, while the error per band was as follows: 9.420 in band 2, 11.601 in band 3, 7.152 in band 4, 1.404 in band 5, 4.338 in band 6, and 2.739 in band 7.

Chapter 6
Fraction Images Applications

Abstract Examples of fraction image application are given for mapping projects of large areas of the Earth's surface such as the operational PRODES project by INPE, which aims to calculate rates of deforestation in the Brazilian Amazon.

Keywords PRODES · Vegetation mapping · DETER · Deforestation estimation

This chapter aims to present some examples of the use of fraction images derived from the linear spectral mixture model in the environmental monitoring projects. The fraction images, due to their characteristics described above (data reduction and enhancement), have contributed to the development of large-scale projects that use a large amount of images.

6.1 Monitoring of Deforestation

The deforestation in the Legal Amazon (AML) has been a concern of various governmental and nongovernmental organizations, especially during the last three decades (Moran 1981; Skole and Tucker 1993). Although there is not a long history of human occupation in AML, almost 90% of deforestation for pasture and agriculture occurred between 1970 and 1988, as indicated by the estimates based on satellite images (Skole et al. 1994).

Historically, the Brazilian territory was occupied along the coastline, with the majority of its population concentrated in this region. In an attempt to change this pattern of occupancy by increasing settlement in the interior of the country, the federal capitol was transferred from the coast, in Rio de Janeiro, to the central region, in Brasilia, in mid-1950 (Mahar 1988). This policy of occupation required infrastructure investments to connect Brasília with the other regions of the country. The construction of the Belém-Brasília highway (BR-010), in 1958, was the dominant factor that triggered the main activities of deforestation in AML (Moran et al. 1994; Nepstad et al. 1997). Subsequent events, such as the construction of the

Y. E. Shimabukuro, F. J. Ponzoni, *Spectral Mixture for Remote Sensing*,
Springer Remote Sensing/Photogrammetry,
https://doi.org/10.1007/978-3-030-02017-0_6

BR-364, crossing the states of Mato Grosso, Rondônia, and Acre, and the PA-150, in the state of Pará, encouraged further deforestation activities, converting forests into pasture and agriculture areas (Moran 1993).

To introduce the governance in AML, the "Superintendência do Desenvolvimento da Amazônia" (SUDAM) and the "Banco da Amazônia" (BASA) were created in 1966. Small producers have been financed to encourage investment in projects of agriculture (Moran et al. 1994). Large producers were also financed through tax incentives in trade for converting forest into pasture areas (Moran 1993). The incentives granted to the large producers were the main causes of deforestation; the small producers have had a smaller impact on deforestation due to the practice of comparatively smaller dimension of the subsistence agriculture (Fearnside 1993).

Other activities with high economic value, such as mining and selective exploitation of timber, also contributed to the deforestation in AML (Cochrane et al. 1999). The areas of deforestation in Brazilian Legal Amazon have been concentrated in the so-called arc of deforestation, located in the southern and eastern parts of AML, from Acre to the Maranhão states (Cochrane et al. 1999; Achard et al. 2002).

6.1.1 The Brazilian Legal Amazon Monitoring Program

Since 1973 Brazil has access to the remote sensing images from Landsat satellite series that allow quantifying the extent of natural resources and the changing of the Amazon region. Based on availability of these images, the Brazilian government started monitoring the Amazon rainforest for quantifying deforestation areas at several years' intervals.

The Brazilian government has conducted annual monitoring of the Amazon forest since 1988 using images generated by the Landsat program through the PRODES (Monitoring of Brazilian Amazon Forest by Satellite) project conducted by the National Institute for Space Research (INPE). It is the largest remote sensing project in the world for monitoring deforestation activity in tropical forests aiming to assess all deforested areas within the five million square kilometers of the AML, which is covered by approximately 229 TM/Landsat 5 scenes (Fig. 6.1).

This project estimated approximately 750,000 km^2 of deforestation in the AML until 2010, a total which accounts for approximately 17% of the original forest extension. These data showed annual rates of deforestation which varied significantly in response to political, economic, and financial conditions of the country as well as foreign market demands.

The PRODES information are based mainly on images of sensors with medium spatial resolution (30 m), such as those generated by the Landsat program, and with relatively low temporal resolution (16 days frequency of acquisition), allowing the annual monitoring of the deforestation. A faster update of forest change is not possible with these images due to the low frequency of acquisition of cloud-free images, which is a serious problem for the Amazon region, limiting the number of viable observations of the ground surface. This fact prevents the Brazilian government and

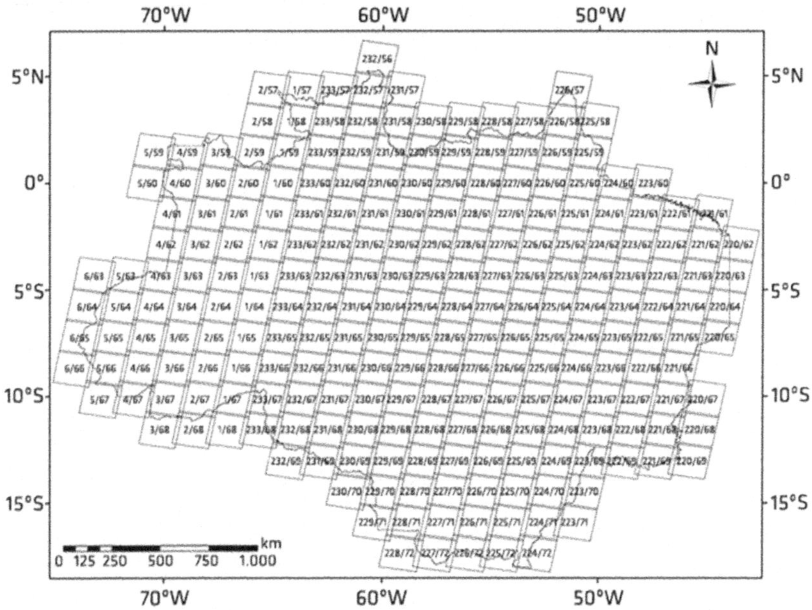

Fig. 6.1 Brazilian Legal Amazon covered by 229 TM or ETM + Landsat images for the estimation of annual deforestation. (Source: INPE (2002))

environment control agencies to make quick and appropriate interventions to stop the activities of illegal deforestation.

Then to monitor the deforestation in near real time is possible using the almost daily images acquired by MODIS sensor on Terra and Aqua platforms. Thus, through the project DETER (near real-time deforestation detection), a new methodology based on MODIS images was developed for the rapid detection of deforestation in the Amazon (Shimabukuro et al. 2006). While the MODIS is a moderate spatial resolution sensor and does not generate viable images to estimate the extension of deforested areas, the MODIS data can be valuable as indicators of changes or as the alarm product for the service of management and control polices of the ground surface.

The following sections show an overview of the digital PRODES and DETER projects for monitoring deforestation activities in the Brazilian Legal Amazon yearly and monthly, respectively. First, the deforestation history and the description of methodology developed at INPE for monitoring based on geographic information system (GIS) and remote sensing image processing are presented. Then, it is shown the important contribution of the fraction images derived from the linear spectral mixture model to make the digital PRODES and DETER viable, which allowed the releasing of the deforestation data in transparent mode to the national and international community. The results provide invaluable contribution to the decision-makers in establishing public policies and strengthen environmental governance in critical ecosystems of the Brazilian Legal Amazon.

6.1.2 The Digital PRODES Project

Since the late 1970, INPE has performed evaluations of deforestation in AML using remote sensing images. These assessments were made in conjunction with the former Brazilian Institute for Forest Development (IBDF), which was later incorporated into the Brazilian Institute of Environment and Renewable Natural Resources (IBAMA). The first assessment was carried out with the use of images acquired by the MSS sensor, with four spectral bands and spatial resolution of ~80 m, onboard of Landsat 1, 2, and 3 satellites, during 1973–1975 and 1975–1978 periods, employing visual interpretation techniques (Tardin et al. 1980).

From 1988 onward, the annual deforestation assessments were provided to the entire AML using the TM sensor, six spectral bands with spatial resolution of 30 m, on board the Landsat 5 satellite, with improved quality of mapping due to its better spatial and spectral resolutions when compared with the MSS data. The methodology applied to map deforested areas was based on visual interpretation of color composites (R5 G4 B3) of TM images in hard-copy format on a scale of 1:250,000. The polygons interpreted visually of the deforested areas were added together to calculate the total deforested areas for each state of the AML and presented in table format. This method, known as analogic PRODES, was performed until 2003.

At the end of 1990 decade, a semi-automatized methodology using the fraction images began to be developed and was named the digital PRODES (Shimabukuro et al. 1998). The digital PRODES project is an automation of the activities performed in the project PRODES, which was based on analogical data since the 1970 decade.

According to PRODES the deforestation in the Brazilian AML is shown by a mask of intact forest that is updated annually through the identification of new deforestation event excluding the non-forest vegetation areas and identifying other dynamic changes, as clear-cut of secondary regeneration areas. Landsat TM images or images from similar sensors are selected for July, August, and September period, which is within the local dry season period on the arc of deforestation and represents an atmospheric window when cloud-free images are normally available. These images are geometric corrected using the technique of sampling the nearest neighbor to the UTM projection resulting in a cartographic product with internal error of 50 m. The spectral bands TM 3 (red), TM 4 (NIR), and TM 5 (MIR) are used to generate the fraction images. The legend for the maps contains the following classes: forest, non-forest (shrub savanna (cerrado), grassy savanna (campo limpo de cerrado), campinarana etc.), deforestation accumulated of previous years, deforestation of the analyzed year, hydrography, and cloud.

The digital PRODES consists of the following methodological steps: (1) generation of vegetation, soil, and shade/water fraction images; (2) image segmentation based on the growing region algorithm; (3) image classification based on unsupervised classifier; (4) mapping classes based on the following legend – forest, non-forest (vegetation that is not characterized by a forest structure), deforestation (clearing accumulated up to the previous year), hydrography, and clouds; and (5) edition of the classified map based on visual interpretation to minimize the omission and commission errors of automatic classification in order to produce the final map of deforestation in digital format. PRODES products are available on the official INPE website.

A linear spectral mixture model is used to produce vegetation, soil, and shade/water fraction images applied to the spectral bands of Landsat TM (Shimabukuro and Smith 1991) (Fig. 6.2). This method reduces the dimensionality of data and highlights the specific targets of interest. The vegetation fraction image highlights the areas of vegetation cover, the soil fraction image highlights the uncovered soil, and the shade/water fraction image highlights the areas of water bodies and the burned areas. The shade/water fraction image was used to characterize the total area deforested before 1997 in AML according to the methodology proposed by Shimabukuro et al. (1998). Subsequently, the deforested areas were accumulated by

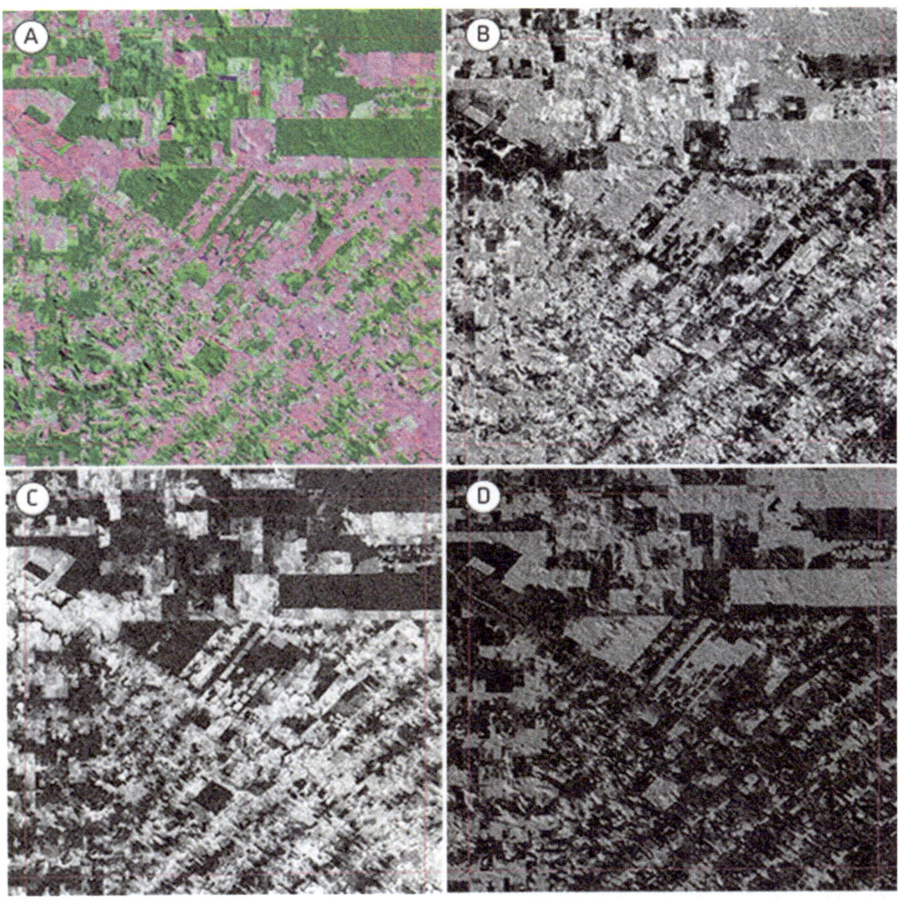

Fig. 6.2 (**a**) Landsat TM image (R5 G4 B3) and fraction images (**b**) vegetation, (**c**) soil, and (**d**) shade/water

the year 2000. From there, the soil fraction image, which highlights the areas without vegetation cover, was used to classify the annual increment of deforestation based on the high contrast between forested and deforested areas in the following years. The digital PRODES, recognized nationally and internationally, allowed INPE to put at the disposal of the general community information of deforested areas in the AML.

The linear spectral mixture model used was:

$$r_i = a\text{vege}_i + b\text{soil}_i + c(\text{water or shade})_i + e_i \quad (6.1)$$

where:

r_i = spectral response for the pixel in the band i of Landsat TM image
a, b, and c = proportion of vegetation, soil, and shade/water in each pixel
vege_i, soil_i, and $(\text{water or shade})_i$ = spectral responses of each component
e_i = error term for each band of Landsat TM image

TM 3, TM 4, and TM 5 bands are used to form a system of linear equations that can be solved by any algorithm developed, for example, weighted least squares (WLS), described earlier. The resulting fraction images were resampled to 60 m to minimize computer processing time and disk space, with no loss of information compatible with the map scale of the final product of 1:250,000.

The next step is the application of a digital image processing technique called image segmentation, which is based on grouping data into contiguous regions with similar spectral characteristics. Two thresholds are required to perform the image segmentation: (a) similarity, which is the minimum value defined by the user to be considered as similar and to form a region, and (b) area, which is the minimum size, in number of pixels, for the region to be individualized (Fig. 6.3). The unsupervised classification method (ISOSEG) is used to classify the segmented fraction images segmented and employs the statistical attributes (mean and covariance matrix) derived from polygons generated by image segmentation (Fig. 6.4).

After the unsupervised classification, it is necessary to check the resulting maps, according to the previously established legend of the PRODES project (Fig. 6.5). Next, the task of editing the map is executed by interpreters using interactive image-editing tools (Fig. 6.6). Omission and commission errors identified by the interpreters are corrected manually in order to improve the classification result.

Then the images individually classified are mosaicked to generate the final maps for each state and for the entire AML (Fig. 6.7). To the mosaic of the states, the spatial resolution is maintained at 60 m, and the scale for presentation is 1:500,000, while for the AML the spatial resolution is resampled to 120 m, and the scale for presentation is 1:2,500,000, due to the large amount of information.

However, the deforestation information provided by the PRODES was not sufficient for the needs of most frequent surveillance of various agencies of the Brazilian government. Therefore, the DETER project was developed, based on images from the MODIS sensor, with high temporal resolution, to provide geospatial information from deforestation activities in near real time.

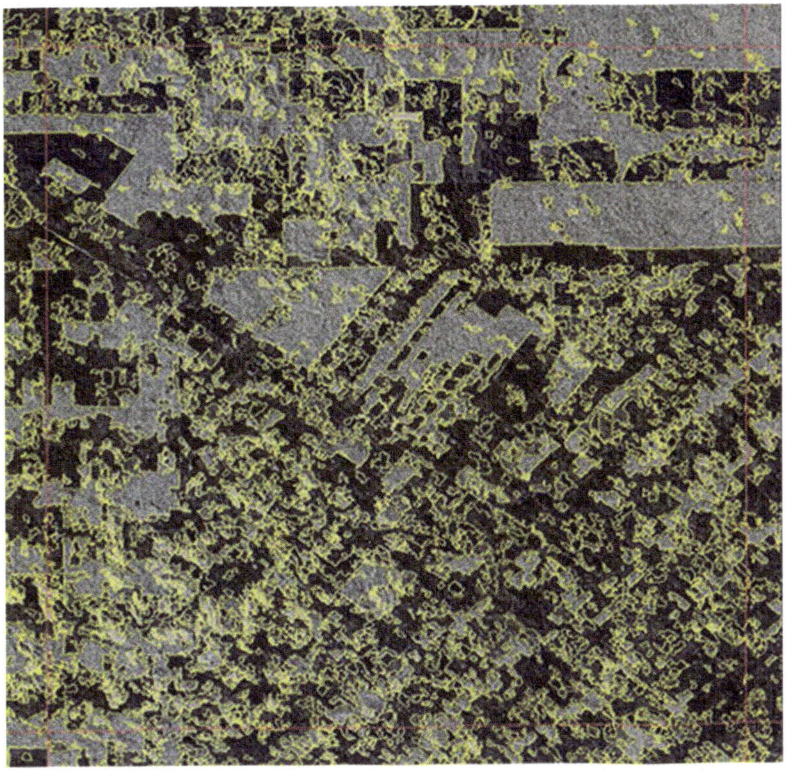

Fig. 6.3 Segmented shade/water fraction image

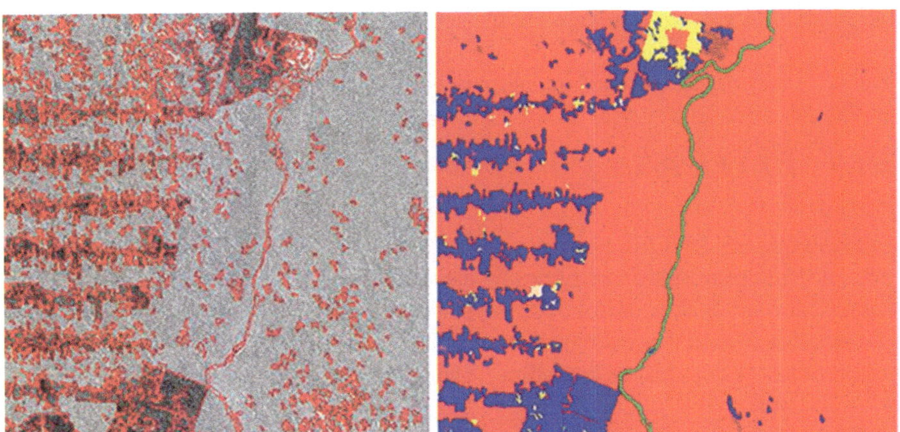

Fig. 6.4 Segmented and classified image using the unsupervised classifier

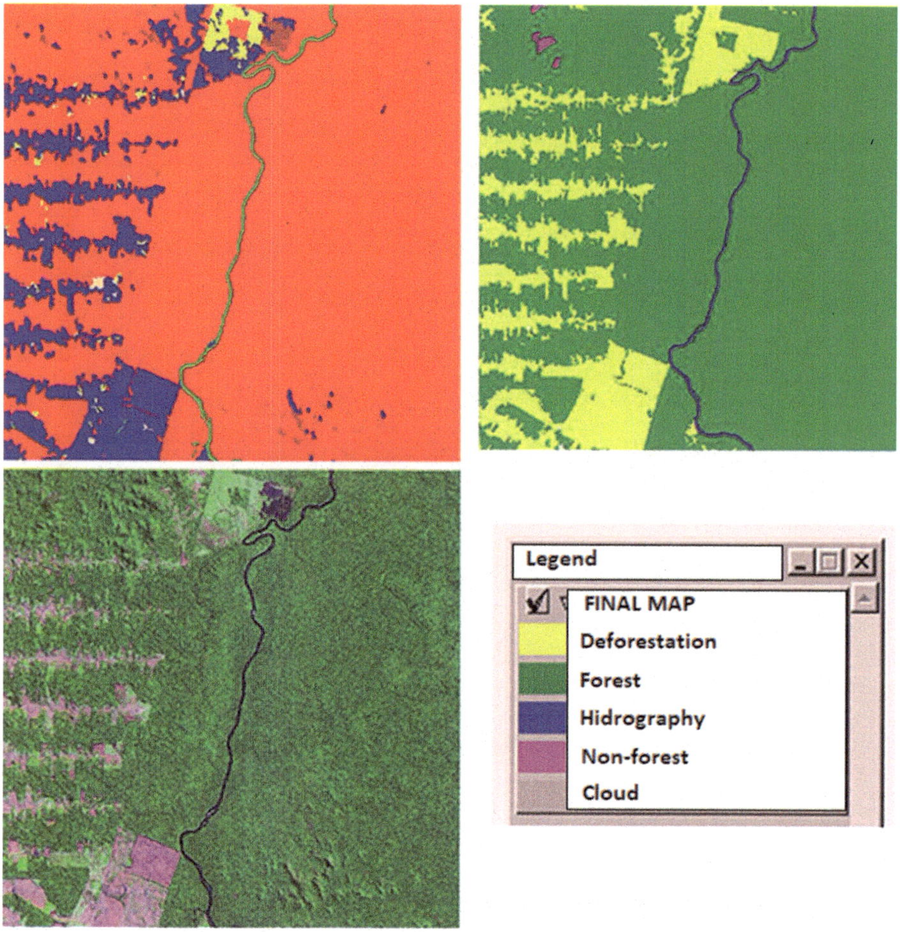

Fig. 6.5 Thematic classes according to the previously established legend of the PRODES project

6.1.3 DETER Project

From 2004, the DETER project was implemented in order to provide real-time detection of deforestation activities to support the federal government's action plan for the prevention and control of deforestation in the Brazilian AML. The procedure mimics the methodology adopted in the PRODES project but is intended to detect deforestation activities in real time, exploring the high temporal resolution of MODIS sensor.

The first step in the method of DETER project is "to mask" the intact forest based on the evaluation of PRODES in the previous year. The map of intact forest is used as a reference for identifying new real-time deforestation events throughout the analyzed year. The monitoring activity with MODIS images starts in January but

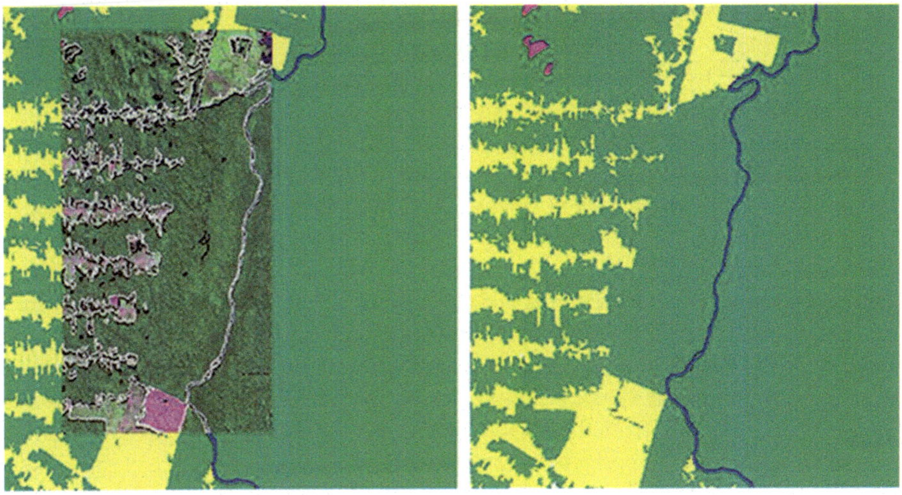

Fig. 6.6 Image edition to generate the final map

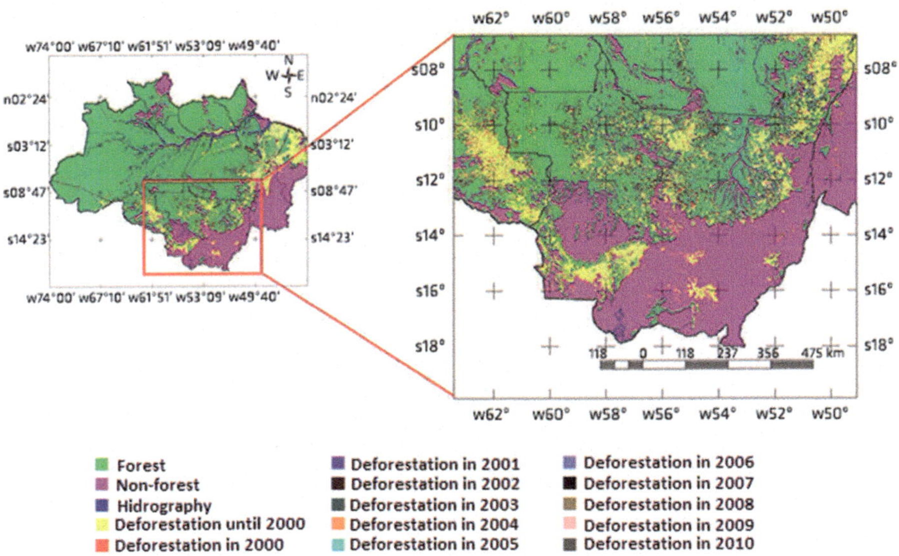

Fig. 6.7 Digital PRODES thematic map showing the deforested areas up to 2000 (in yellow), non-forest areas (in magenta), and annual deforestation of 2001–2010 according to the legend

becomes more effective after March, when a larger number of MODIS images are available due to less cloud cover in AML. In addition, during the rainy season, from November to March, there is not much deforestation expected to happen.

Daily MODIS images (surface reflectance-MOD09) used to identify sources of deforestation are selected based on two criteria: (a) amount of cloud coverage and

(b) track inside the Zenithal angle of the sensor smaller than 35° (~1400 km). The amount of cloud cover is initially evaluated in accordance with the quick-look images followed by more detailed analysis with actual spatial resolution of MODIS images. The AML is covered by 12 MODIS tiles (V09 to V11 and H10 to H13).

The images of MOD09 product are in sinusoidal projection (WGS84 datum), and the bands are projected for the geographic coordinate system with the same datum and converted from HDF (hierarchical data format) to GeoTIFF in order to download the images directly to SPRING software for image processing proposes. High geometric quality of MODIS products ensures the viability of the project, because it is crucial to detect points of deforestation within the MODIS pixel size.

From the set of seven bands of MOD09 product, band 1 (red), band 2 (NIR), and band 6 (MIR) are used to generate the vegetation, soil, and shade/water fraction images applying linear spectral mixture model (see Sect. 6.1.2), as can be seen in Fig. 6.8 for the period from 22 April to 7 May 2004. In this figure, the soil fraction image facilitates the mapping of deforested areas.

The soil fraction images are then segmented, classified, mapped, and eventually edited by interpreter following the same methods used in the digital PRODES project. Figure 6.9 shows the steps of the DETER project, i.e., the deforested areas classified in the soil fraction images are superimposed on the forest mask, highlighting the areas of new deforestation (red color).

This procedure is performed for each daily MODIS image acquired for the Brazilian AML. The results of the deforestation activities detected by DETER can be accumulated for different intervals such as weekly, biweekly, and monthly and are available in digital format on the project website of INPE. Figure 6.10 displays the products available for the year 2004 for this project.

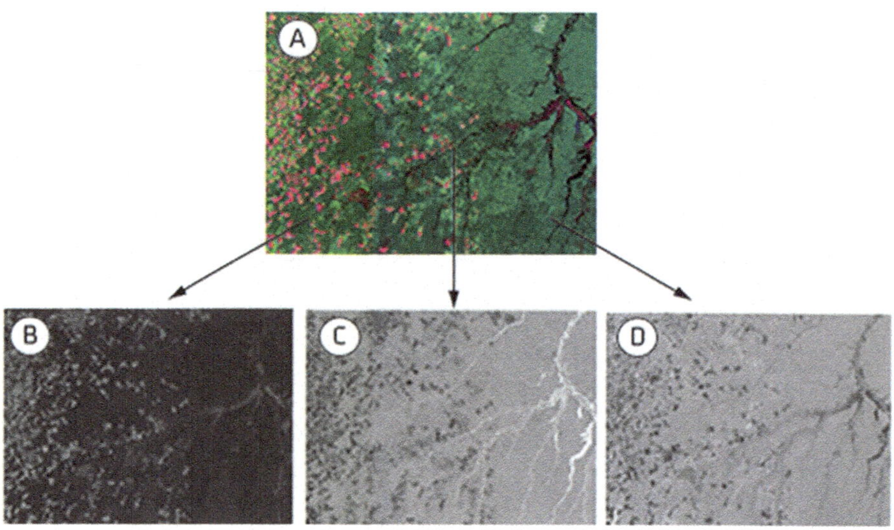

Fig. 6.8 (**a**) MODIS image corresponding to the period of 22 April to 7 May 2004 and fraction images (**b**) soil, (**c**) shade/water, and (**d**) vegetation

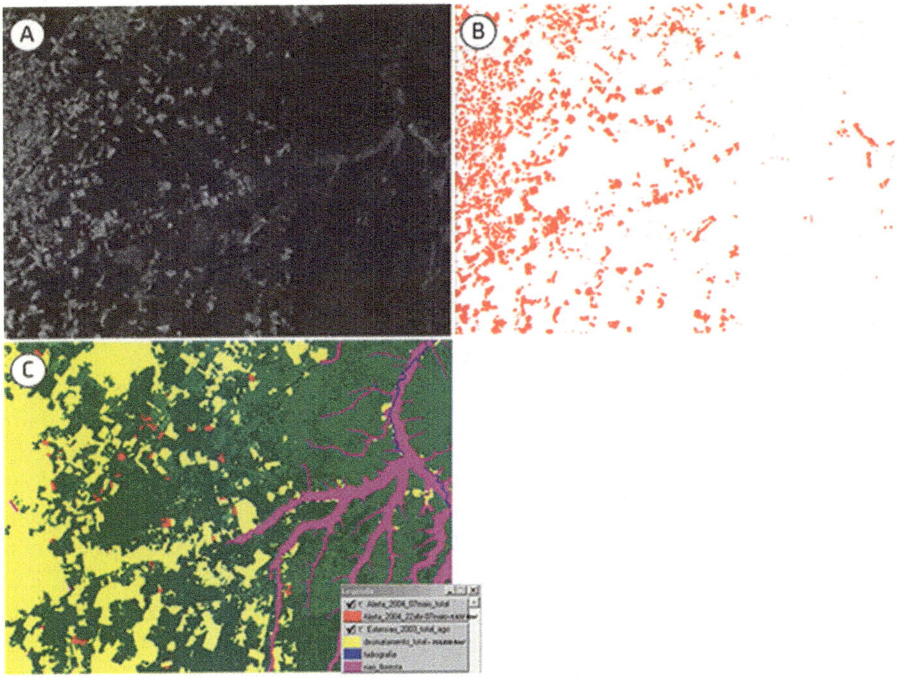

Fig. 6.9 (**a**) Soil fraction image (mosaic of 22 April to 7 May 2004); (**b**) result of the classification of the soil fraction image; (**c**) classification of the MODIS image (mosaic of 22 April to 7 May 2004) – total extension: August 2003 + change until 7 May 2004

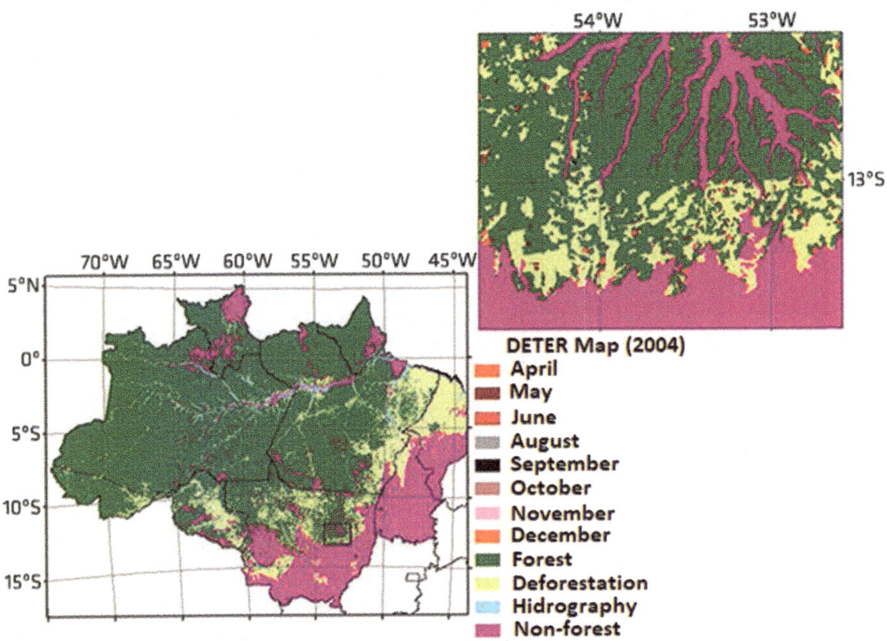

Fig. 6.10 Result of monitoring of DETER project showing deforestation activities monthly detected during the year 2004

6.2 Burned Areas Mapping

Beyond the capacity of the MODIS images of serving for the real-time detection of the deforested areas (DETER project), through them it is possible also to identify other types of anthropogenic action on the forest cover, as is the case of burning activities. This can be seen with the results obtained in the state of Acre using MOD09 product acquired on 5, 12, and 21 September 2005, composed by the spectral bands of red (centered at 640 nm), near infrared (858 nm), and middle infrared (1640 nm) (Shimabukuro et al. 2009).

For the initial interpretative phase of burned areas using MODIS images, it was used the information from the PROARCO project, which presents the daily monitoring of hot pixels. The burned areas mapping project makes possible to estimate the spatial distribution, the degree of incidence and the spatial direction of the fire. That information can be compared to thematic information allowing to determine the vegetation type that is being affected. For validation of the burned areas map using MODIS data, it was used the images of better spatial resolution, case of Landsat TM and the CBERS-2/CCD products, acquired on 13 and 12 October 2005, respectively, in addition to field information in the considered period. In Fig. 6.11 it is presented the flowchart of procedures performed on the work.

Figure 6.12 shows the color composite (R6 G2 B1) and also the individual vegetation, soil, and shade/water fraction images for the MODIS image acquired on 12

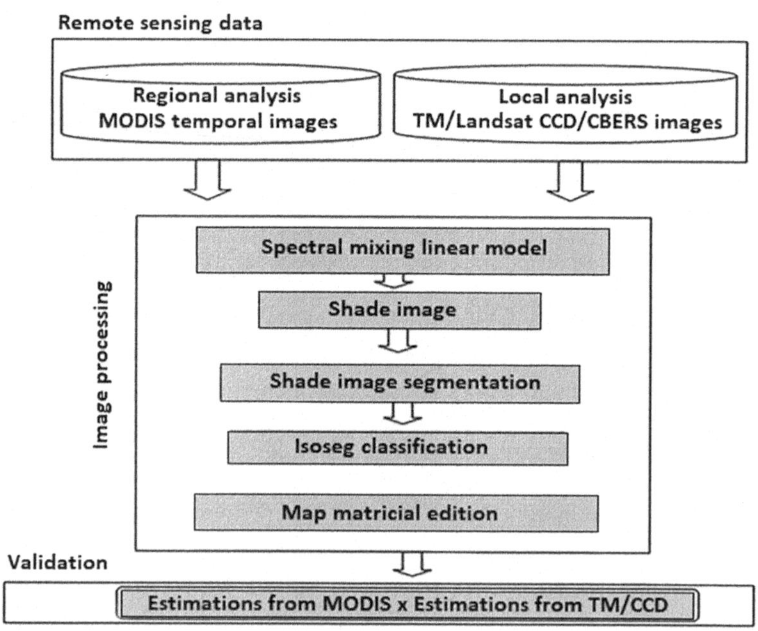

Fig. 6.11 Flowchart of the methodology of burned areas mapping

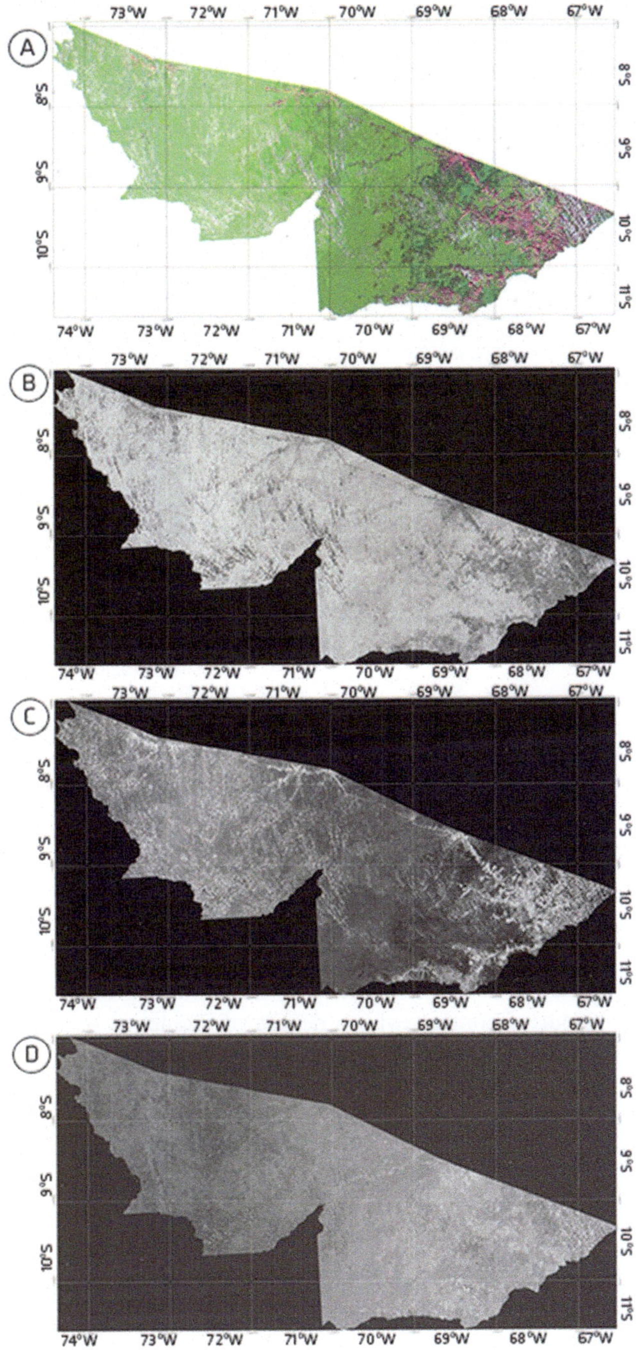

Fig. 6.12 MODIS image of 12 October 2005 of the state of Acre: (**a**) color composite (R6 G2 B1) and fraction images (**b**) vegetation, (**c**) soil, and (**d**) shade/water

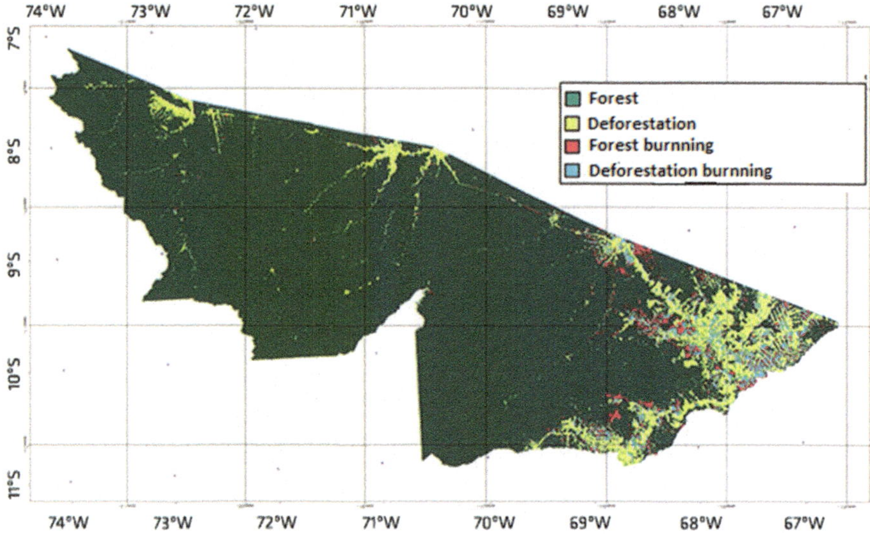

Fig. 6.13 Burned areas in the state of Acre identified in the MODIS images acquired in the year 2005

October 2005. The burned areas can be identified as gray levels higher than the other targets in the shade/water fraction image, facilitating their discrimination.

The results of the analysis of multitemporal shade/water fraction images derived from MODIS data indicated the occurrence of 6500 km^2 of the area burned in the state of Acre (Fig. 6.13). Of this total, 3700 km^2 correspond to the previously deforested areas, where burning activity serves as a traditional practice of cleaning the ground for the implementation of agricultural crops or new pastures or even as a practice of improvement of grazing. The other 2800 km^2 correspond to the areas of forest fires, with the forest cover degraded by fire, in both ground level and canopy level, in burning out of control, whose degree of incidence is determined by the wind action, availability of dry material in the interior of the forest, and occurrence of certain species more susceptible to fire.

Concerned with the level of emissions resulting from burning and their impacts on air quality breathed by the population of the state of Acre, Brown et al. (2006) estimated using TM/Landsat 5 and CCD/CBERS-2 images and a detailed fieldwork; more than 2670 km^2 of primary forests were affected by fires in the southeast of that state in 2005. Such results show the consistency of assessments with MODIS images, qualifying them as an important source of information for mapping burned areas on a regional scale.

6.3 Selective Logging Detection

The selective logging of species of high commercial value is a practice used in areas of the Amazon rainforest requiring several studies for their detection and measurement and the evaluation of the impacts on intact forests (Asner et al. 2005; Grogan et al. 2008; Matricardi et al. 2010; Shimabukuro et al. 2014).

Selective logging is characterized by the opening of stocking areas and exploration tracks. In this way, with the use of medium spatial resolution images (e.g., TM/Landsat 5), it is possible to detect these areas using the soil fraction image generated by linear spectral mixture model. Figure 6.14a shows the soil fraction image highlighting the deforested and selective logging areas, which can be classified as shown in Fig. 6.14b.

6.4 Land Use and Land Cover Mapping

The vegetation, soil, and shade/water fraction images have been used for mapping the land use and the land cover. The following example is the mapping of the Mato Grosso state using multitemporal data of MODIS/Terra. Figure 6.15a, b presents the color composite of MODIS images acquired in the months of January and August 2002, respectively, showing the landscape change during the rainy and drought periods in the Mato Grosso state. In the image from August, it is observed the contrast between areas occupied by bare soils and covered by very sparse vegetation (cerrado areas and deforested areas) and areas occupied by dense vegetation (forest-covered areas). On the other hand, in the January image, it is observed the agriculture areas, mainly soybeans, and flooded areas. Then, the fraction images of the MODIS

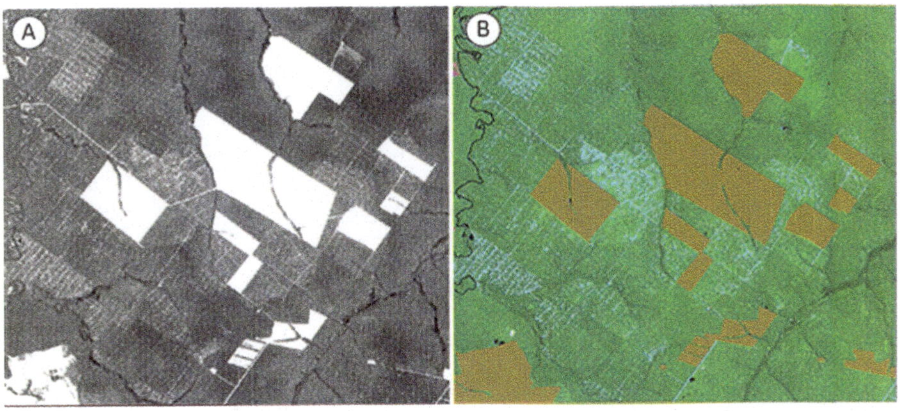

Fig. 6.14 (**a**) Soil fraction image derived from a TM image on an area in the state of Mato Grosso, highlighting the deforested areas (clear-cut) and the areas of selective logging; (**b**) classification of selective logging areas (light blue) and deforested areas (brown)

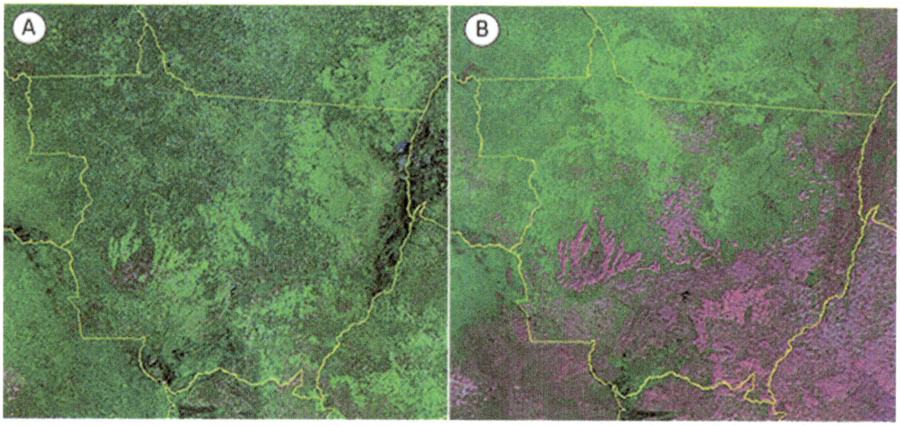

Fig. 6.15 Color composites of MODIS images of the state of Mato Grosso acquired in (**a**) January 2002 and (**b**) August 2002

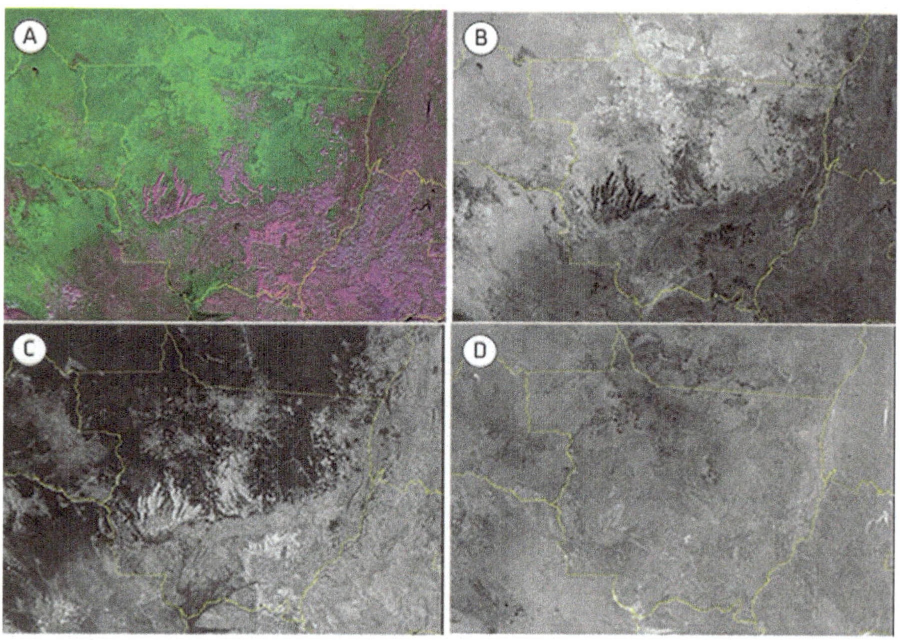

Fig. 6.16 MODIS fraction images of August 2002 of Mato Grosso state: (**a**) color composite, (**b**) vegetation, (**c**) soil, and (**d**) shade/water

data acquired during the year can be used to map the land use and land cover areas. Therefore, the fraction images are very useful to reduce the data volume to be analyzed, in addition to enhance the land cover classes of interest.

Figure 6.16 displays the fraction images derived from the image acquired in the month of August 2002, facilitating the discrimination between vegetation and non-vegetation areas. It is also possible to differentiate the deforested areas and cer-

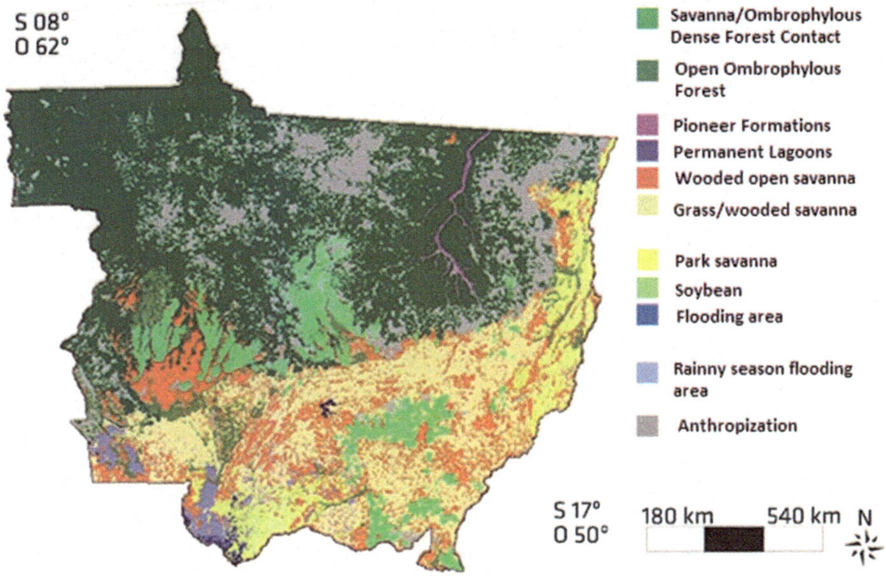

Fig. 6.17 Map of land use and land cover of Mato Grosso state obtained based on the MODIS images. (Source: Anderson (2004))

rado areas that are without vegetation cover during that period of the year, as well as the water bodies.

Similarly, the fraction images of the month of January 2002, not presented here, highlight the areas of agriculture, mainly soybeans, and different types of savanna (cerrado) and wetlands. Anderson (2004), combining the fraction images of MODIS images acquired during the periods of the year, mapped land use classes and the land cover of the state of Mato Grosso in 2002 as shown in Fig. 6.17.

Chapter 7
Final Considerations

Abstract This chapter presents some final thoughts about the book.

Keywords LSMM objectives · Future applications

The spectral mixture can be linear and nonlinear. The linear model was discussed because of the facility of implementation with very satisfactory results.

The linear spectral mixture model is a technique of data transformation of remote sensing data, i.e., converts the spectral information into physical proportion information of the components (endmembers) within the pixel. This information of proportion of the components is represented in new images called fraction images. In that way, the linear spectral mixture model is a data reduction technique, and in addition it enhances the information of these components within the image pixel. It is not a thematic classifier, but provides useful information of fraction images for a variety of applications in several areas.

In general, these endmembers are vegetation, soil, and shade/water elements present on the ground. The vegetation fraction image presents similar information of vegetation indices such as NDVI, SAVI, and EVI, highlighting the vegetation cover areas, while the soil fraction image highlights the areas without vegetation cover, and the shade/water fraction image highlights the water bodies and the burned areas.

The soil and shade/water fraction images were important for automating the PRODES project, which was done through the digital PRODES project, providing the estimate of deforested areas and the map of spatial distribution of these areas.

Hopefully, at the end of this book, we have contributed to the provision of useful information to the deepest reflections for those who intend to use the fraction images derived from linear spectral mixture model in the development of their works.

Y. E. Shimabukuro, F. J. Ponzoni, *Spectral Mixture for Remote Sensing*,
Springer Remote Sensing/Photogrammetry,
https://doi.org/10.1007/978-3-030-02017-0_7

References

Achard, F., Eva, H. D., Stibig, H. J., Mayaux, P., Gallego, J., Richards, T., & Malingreau, J. P. (2002). Determination of deforestation rates of the world's humid tropical forests. *Science, 297*, 999–1002.

Adams, J. B., & Adams, J. D. (1984). Geologic mapping using Landsat MSS and TM images: Removing vegetation by modeling spectral mixtures – remote sensing for exploration geology. In *Proceedings of the third thematic conference: International symposium on remote sensing of environment, Colorado* (pp. 16–19). Colorado Springs: ERIM.

Adams, J. B., Smith, M. O., & Johnson, P. E. (1986). Spectral mixture modeling: A new analysis of rock and soil types at the Viking Lander 1 site. *Journal of Geophysical Research, 91*(B8), 8098–8112.

Alcântara, E. H., Barbosa, C. C. F., Stech, J. L., Novo, E. M. M., & Shimabukuro, Y. E. (2009). Improving the spectral unmixing algorithm to map water turbidity distributions. *Environmental Modelling & Software, 24*, 1051–1061.

Anderson, L. O. (2004). *Classificação e monitoramento da cobertura vegetal de Mato Grosso utilizando dados multitemporais do sensor MODIS.* 247 f. Dissertação (Mestrado em Sensoriamento Remoto) – Instituto Nacional de Pesquisas Espaciais, São José dos Campos. INPE-12290-TDI/986.

Asner, G. P., Knapp, D. E., Broadbent, E. N., Oliveira, P. J. C., Keller, M., & Silva, J. N. (2005). Selective logging in the Brazilian Amazon. *Science, 310*, 480–482.

Atkinson, P., Cutler, M., & Lewis, H. (1997). Mapping subpixel proportional land cover with AVHRR imagery. *International Journal of Remote Sensing, 18*, 917–935.

Bastin, L. (1997). Comparison of fuzzy c-means classification, linear mixture modeling and MLC probabilities as tools for unmixing coarse pixels. *International Journal of Remote Sensing, 18*, 3629–3648.

Boardman, J. W.(1989). Inversion of imaging spectrometry data using singular value decomposition. In: *Proceedings of the 12th Canadian symposium on remote sensing*. (Vol. 4, pp. 2069–2072). Vancouver, Canadá: IEEE IGARSS.

Brown, I. F., Schroeder, W., Setzer, A., Maldonado, M., Pantoja, N., Duarte, A., & Marengo, J. (2006). Monitoring fires in southwestern Amazonian rain forests. *EOS Transactions, 87*(26), 253–259.

Burden, R. L., Faires, J. D., & Reynolds, A. C. (1981). *Numerical analysis* (2nd ed.). Boston: Prindle, Weber and Schmidt.

Chander, G., Markham, B. L., & Barsi, J. A. (2007). Revised Landsat 5 thematic mapper radiometric calibration. *IEEE Transactions on Geoscience and Remote Sensing Letters, 4*(3), 490–495.

Y. E. Shimabukuro, F. J. Ponzoni, *Spectral Mixture for Remote Sensing*, Springer Remote Sensing/Photogrammetry, https://doi.org/10.1007/978-3-030-02017-0

Chander, G., Haque, O., Micijevic, E., & Barsi, J. A. (2010). A procedure for radiometric recalibration of Landsat 5 TM reflective-band data. *IEEE Transactions on Geoscience and Remote Sensing, 8*(1), 556–574.

Cochrane, M. A., Alencar, A., Schulze, M. D., Souza, C. M., Jr., Nepstad, D. C., Lefebvre, P., & Davidson, E. A. (1999). Positive feedbacks on the fire dynamic of closed canopy tropical forests. *Science, 284*, 1832–1835.

Conte, S. D., & De Boor, C. (1980). *Elementary numerical analysis: An algorithmic approach* (p. 445). New York, USA: McGraw-Hill.

Detchmendy, D. M., & Pace, W. H. (1972). A model spectral signature variability for mixtures. In F. Shahrokhi (Ed.), *Remote sensing of Earth resources* (Vol. 1, pp. 596–620). Tullahoma: The University of Tennessee.

Dietz, A., Wohner, C., & Kuenzer, C. (2012). European snow cover characteristics between 2000 and 2011 derived from improved MODIS daily snow cover products. *International Journal of Remote Sensing, 4*, 2432–2454.

Dietz, A., Kuenzer, C., & Conrad, C. (2013). Snow cover variability in Central Asia between 2000 and 2011 derived from improved MODIS daily snow cover products. *International Journal of Remote Sensing, 34*(11), 3879–3902.

Fearnside, P. M. (1993). Deforestation in Brazilian Amazonia: The effect of population and land tenure. *Ambio, 22*(8), 537–545.

Foody, G. M., Lucas, R. M., Curran, P. J., & Honzak, M. (1997). Non-linear mixture modeling without end-members using an artificial neural network. *International Journal of Remote Sensing, 18*, 937–953.

García-Haro, F., Sommner, S., & Kemper, T. (2005). A new tool for variable multiple endmember spectral mixture analysis (VMESMA). *International Journal of Remote Sensing, 26*, 2135–2162.

Gessner, U., Machwitz, M., Esch, T., Tillack, A., Naeimi, V., Kuenzer, C., & Dech, S. (2015). Multi-sensor mapping of West African land cover using MODIS, ASAR and TanDEM-X/TerraSAR-X data. *Journal of Remote Sensing Environment, 164*, 282–297.

Gilabert, M. A., Conese, C., & Maselli, F. (1994). An atmospheric correction method for the automatic retrieval of surface reflectances from TM images. *International Journal of Remote Sensing, 15*(10), 2065–2086.

Grogan, J., Jennings, S. B., Landis, R. M., Schulze, M., Baima, A. M. V., Lopes, J. C. A., Norghauer, J. M., Oliveira, L. R., Pantoja, F., Pinto, D., Silva, J. N. M., Vidal, E., & Zimmerman, B. L. (2008). What loggers leave behind: Impacts on big-leaf mahogany (*Swietenia macrophylla*) commercial populations and potential for post-logging recovery in the Brazilian Amazon. *Forest Ecology and Management, 255*, 269–281.

Heimes, F. J. (1977). *Effects of scene proportions on spectral reflectance in Lodgepole pine*. Dissertation (Master of Science) – Colorado State University, Fort Collins.

Horwitz, H. M., Nalepka, R. F., Ryde, P. D., & Morgenstern, J. P. (1971). Estimating the proportions of objects within a single resolution element of a multispectral scanner. In *Proceedings of the 7th international symposium on remote sensing of environment, May 7–21, 1971* (pp. 1307–1320). Ann Arbor: Willow Run Laboratories.

INPE – Instituto Nacional de Pesquisas Espaciais. (2002). *Monitoring of the Brazilian Amazonian forest by satellite, 2000-2001* (p. 21). São José dos Campos, SP, Brasil: Instituto Nacional de Pesquisas Espaciais.

Klein, I., Gessner, U., & Kuenzer, C. (2012). Regional land cover mapping in Central Asia using MODIS time series. *Applied Geography, 35*, 1–16.

Klein, I., Dietz, A., Gessner, U., Dech, S., & Kuenzer, C. (2015). Results of the global waterpack: A novel product to assess inland water body dynamics on a daily basis. *Remote Sensing Letters, 6*(1), 78–87.

Kuenzer, C., Klein, I., Ullmann, T., Foufoula-Georgiou, E., Baumhauer, R., & Dech, S. (2015). Remote sensing of river delta inundation: Exploiting the potential of coarse spatial resolution, temporally-dense MODIS time series. *Remote Sensing, 7*, 8516–8542.

Leinenkugel, P., Wolters, M., Oppelt, N., & Kuenzer, C. (2014). Tree cover and forest cover dynamics in the Mekong Basin from 2001 to 2011. *Remote Sensing of Environment, 158*, 376–392.
Lu, L., Kuenzer, C., Guo, H., Li, Q., Long, T., & Li, X. (2014). A novel land cover classification map based on MODIS time-series in Nanjing, China. *Remote Sensing, 6*, 3387–3408.
Lu, L., Kuenzer, C., Wang, C., Guo, H., & Li, Q. (2015). Evaluation of three MODIS-derived vegetation index time series for dry land vegetation dynamics monitoring. *Remote Sensing, 7*, 7597–7614.
Mahar, D. (1988). *Government policies and deforestation in Brazil's Amazon Region*. Washington, D.C.: World Bank.
Matricardi, E. A. T., Skole, D. L., Pedlowski, M. A., Chomentowski, W., & Fernandes, L. C. (2010). Assessment of tropical forest degradation by selective logging and fire using Landsat imagery. *Remote Sensing of Environment, 114*, 1117–1129.
Moran, E. F. (1981). *Developing the Amazon*. Bloomington: Indiana University Press.
Moran, E. F. (1993). Deforestation and land use in the Brazilian Amazon. *Human Ecology, 21*(1), 1–21.
Moran, E. F., Brondizio, E., Mausel, P., & Wu, Y. (1994). Integrating Amazonian vegetation, land-use, and satellite data. *Bioscience, 44*(5), 329–338.
Nepstad, D. C., Klink, C. A., Uhl, C., Vieira, I. C., Lefebvre, P., Pedlowski, M., Matricardi, E., Negreiros, G., Brown, I. F., Amaral, E., Homma, A., & Walker, R. (1997). Land-use in Amazonia and the Cerrado of Brazil. *Ciência e Cultura Journal of the Brazilian Association for the Advancement of Science, 49*(1/2), 73–86.
Novo, E. M. L. M., & Shimabukuro, Y. E. (1994). Spectral mixture analysis of inland tropical waters. *International Journal of Remote Sensing, 15*(6), 1354–1356.
Pace, W. H., & Detchmendy, D. M. (1973). A fast algorithm for the decomposing of multispectral data into mixtures. In F. Shahrokhi (Ed.), *Remote sensing of Earth resources* (Vol. 2, pp. 831–847). Tullahoma: The University of Tennessee.
Pearson, R. (1973). *Remote multispectral sensing of biomass*. Dissertation (Ph.D.) – Colorado State University, Fort Collins.
Piromal, R. A. S. (2006). *Avaliação do modelo 5-scale para simular valores de reflectância de unidades de paisagem da Floresta Nacional do Tapajós*. 151 f. Dissertação (Mestrado em Sensoriamento Remoto) – São José dos Campos, Inpe. INPE-14645-TDI/205.
Ranson, K. J. (1975). *Computer assisted classification of mixtures with simulated spectral signatures*. Dissertation (Master of Science) – Colorado State University, Fort Collins.
Roberts, D. A., Smith, M., & Adams, J. (1993). Green vegetation, nonphotosynthetic vegetation, and soils in AVIRIS data. *Remote Sensing of Environment, 44*, 255–269.
Rosin, P. (2001). Robust pixel unmixing. *IEEE Transactions on Geoscience and Remote Sensing, 39*, 1978–1983.
Shimabukuro, Y. E. (1987). *Shade images derived from linear mixing models of multispectral measurements of forested areas*. (274 p.) Thesis (Doctor of Phylosophy) – Colorado State University, Fort Collins.
Shimabukuro, Y. E., & Smith, J. A. (1991). The least-squares mixing models to generate fraction images derived from remote sensing multispectral data. *IEEE Transactions on Geoscience and Remote Sensing, 29*, 16–20.
Shimabukuro, Y. E., & Smith, J. A. (1995). Fraction images derived from Landsat TM and MSS data for monitoring reforested areas. *Canadian Journal of Remote Sensing, 21*(1), 67–74.
Shimabukuro, Y. E., Batista, G. T., Mello, E. M. K., Moreira, J. C., & Duarte, V. (1998). Using shade fraction image segmentation to evaluate deforestation in Landsat Thematic Mapper images of the Amazon region. *International Journal of Remote Sensing, 19*(3), 535–541.
Shimabukuro, Y. E., Duarte, V., Anderson, L. O., Valeriano, D. M., Arai, E., & Freitas, R. M. (2006). Near real time detection of deforestation in the Brazilian Amazon using MODIS imagery. *Revista Ambi-Agua, 1*, 37–47.
Shimabukuro, Y. E., Duarte, V., Arai, E., Freitas, R. M., Lima, A., Valeriano, D. M., Brown, I. F., & Maldonado, M. L. R. (2009). Fraction images derived from Terra MODIS data for mapping burnt areas in Brazilian Amazonia. *International Journal of Remote Sensing, 30*, 1537–1546.

Shimabukuro, Y. E., Beuchle, R., Grecchi, R. C., & Achard, F. (2014). Assessment of forest degradation in Brazilian Amazon due to selective logging and fires using time series of fraction images derived from Landsat ETM+ images. *Remote Sensing Letters, 5*, 773–782.

Singer, R. B., & Mccord, T. B. (1979). Mars: Large scale mixing of bright and dark materials and implications for analysis of spectral reflectance. In *Proceedings of the 10th lunar and planetary science conference, Houston, Texas* (pp. 1825–1848). Houston: Lunar and Planetary Institute.

Skole, D., & Tucker, C. (1993). Tropical deforestation and habitat fragmentation in the Amazon: Satellite data from 1978 to 1988. *Science, 260*, 1905–1910.

Skole, D. L., Chomentowski, W. H., Salas, W. A., & Nobre, A. D. (1994). Physical and human dimensions of deforestation in Amazonia. *Bioscience, 44*(5), 14–322.

Slater, P. N. (1980). *Remote sensing, optics and optical systems*. Reading: Addison-Wesley Publishing Company, Advanced Book Program.

Smith, M. O., Johnson, P. E., & Adams, J. B. (1985). Quantitative determination of mineral types and abundances from reflectance spectra using principal component analysis. *Journal of Geophysical Research, 90*(S02), 797–804.

Spiegel, M. R. (1968). *Mathematical handbook of formulas and tables*. New York: McGraw-Hill.

Tardin, A. T., Lee, D. C. L., Santos, R. J. R., Assis, O. R., Barbosa, M. P. S., Moreira, M. L., Pereira, M. T., Silva, D., & Filho, C. P. (1980). Subprojeto desmatamento, Convênio IBDF/CNPq/INPE, Rel. In *Técnico INPE-1649-RPE/103*.

Ustin, S. L., Adams, J. B., Elvidge, C. D., Rejmanek, M., Rock, B. N., Smith, M. O., Thomas, R. W., & Woodward, R. A. (1986). Thematic mapper studies of semiarid shrub communities. *Bioscience, 36*(7), 446–452.

Recommended References

Adams, J. B., & Gillespie, A. R. (2006). *Remote sensing of landscapes with spectral images: A physical modelling approach* (Vol. 378). Cambridge: Cambridge University Press.

Adams, J. B., Smith, M. O., & Gillespie, A. R. (1993). Imaging spectroscopy: Interpretation based on spectral mixture analysis. In C. M. Pieters & P. A. Englert (Eds.), *Remote geochemical analysis: Elemental and mineralogical composition* (Cap. 7, pp. 145–166). New York: Cambridge University Press.

Adams, J. B., Sabol, D. E., Kapos, V., Almeida-Filho, R., Roberts, D. A., Smith, M. O., & Guillespie, A. R. (1995). Classification of multispectral images based on fractions of endmembers: Application to land-cover change in the Brazilian Amazon. *Remote Sensing of Environment, 52*, 137–152.

Aguiar, A. (1991). *Utilização de atributos derivados de proporções de classes dentro de um elemento de resolução de imagem ('pixel') na classificação multiespectral de imagens de sensoriamento remoto*. Dissertação (Mestrado) – Instituto Nacional de Pesquisas Espaciais – INPE, São José dos Campos.

Aguiar, A. P. D., Shimabukuro, Y. E., & Mascarenhas, N. D. A. (1999). Use of synthetic bands derived from mixing models in the multispectral classification of remote sensing images. *International Journal of Remote Sensing, 20*(4), 647–657.

Alcântara, E. H., Stech, J. L., Novo, E. M. M., Shimabukuro, Y. E., & Barbosa, C. C. F. (2008). Turbidity in the Amazon floodplain assessed through a spatial regression model applied to fraction images derived from MODIS/Terra. *IEEE Transactions on Geoscience and Remote Sensing, 46*, 2895–2905.

Anderson, L. O., Shimabukuro, Y. E., & Arai, E. (2005a). Cover: Multitemporal fraction images derived from Terra MODIS data for analysing land cover change over the Amazon region. *International Journal of Remote Sensing, 26*(11), 2251–2257.

Anderson, L. O., Shimabukuro, Y. E., Defries, R. S., & Morton, D. (2005b). Assessment of deforestation in near real time over the Brazilian Amazon using temporal fraction images derived from Terra MODIS. *IEEE Geoscience and Remote Sensing Letters, 2*(3), 315–318.

Anderson, L. O., Aragão, L. E. O. C., Lima, A., & Shimabukuro, Y. E. (2005c). Detecção de cicatrizes de áreas queimadas baseada no modelo linear de mistura espectral e imagens índice de vegetação utilizando dados multitemporais do sensor MODIS/Terra no Estado do Mato Grosso, Amazônia Brasileira. *Acta Amazonica, Manaus, AM, 35*(4), 445–456.

Carreiras, J. M. B., Shimabukuro, Y. E., & Pereira, J. M. C. (2002). Fraction images derived from SPOT-4 VEGETATION data to assess land-cover change over the State of Mato Grosso, Brazil. *International Journal of Remote Sensing, 23*(23), 4979–4983.

Cochrane, M. A., & Souza, C. M. (1998). Linear mixture model classification of burned forest in the Eastern Amazon. *International Journal of Remote Sensing, 19*(17), 3433–3440.

Cross, A., Settle, J. J., Drake, N. A., & Paivinen, R. T. M. (1991). Subpixel measurement of tropical forest cover using AVHRR data. *International Journal of Remote Sensing, 12*(5), 1119–1129.

Ferreira, M. E. (2003). *Análise do modelo linear de mistura espectral na discriminação de fitofisionomias do Parque Nacional de Brasília (Bioma Cerrado)*. 111 f. Dissertação – Instituto de Geociências, Departamento de Geologia Geral e Aplicada da Universidade de Brasília, UnB, Brasília.

Ferreira, M. E., Ferreira, L. G., Sano, E., & Shimabukuro, Y. E. (2007). Spectral linear mixture modelling approaches for land cover mapping of tropical savanna areas in Brazil. *International Journal of Remote Sensing, Grã-Bretanha, 28*(2), 413–429.

García-Haro, F., Gilabert, M., & Meliá, J. (1996). Linear spectral mixture modelling to estimate vegetation amount from optical spectral data. *International Journal of Remote Sensing, 17*(17), 3373–3400.

Haertel, V. F., & Shimabukuro, Y. E. (2005). Spectral linear mixing model in low spatial resolution image data. *IEEE Transactions on Geoscience and Remote Sensing, 43*(11), 2555–2562.

Haertel, V., Shimabukuro, Y. E., & Almeida-Filho, R. (2004). Fraction images in multitemporal change detection. *International Journal of Remote Sensing, 25*(23), 5473–5489.

Hall, F. G., Shimabukuro, Y. E., & Huemmrich, K. F. (1995). Remote sensing of forest biophysical structure in boreal stands of *Picea mariana* using mixture decomposition and geometric reflectance models. *Ecological Applications, 5*, 993–1013.

Hlavka, C., & Spanner, M. (1995). Unmixing AVHRR imagery to assess clearcuts and forest regrowth in Oregon. *IEEE Transactions on Geoscience and Remote Sensing, 33*(3), 788–795.

Holben, B., & Shimabukuro, Y. E. (1993). Linear mixing model applied to coarse spatial resolution data from multispectral satellite sensors. *International Journal of Remote Sensing, 14*(11), 2231–2240.

Jasinski, M., & Eagleson, P. (1990). Estimation of subpixel vegetation cover using red-infrared scattergrams. *IEEE Transactions on Geoscience and Remote Sensing, 28*(2), 253–267.

Kawakubo, F. S. (2010). *Metodologia de classificação de imagens multiespectrais aplicada ao mapeamento do uso da terra e cobertura vegetal na Amazônia*: Exemplo de caso na região de São Félix do Xingu, sul do Pará. 129 f. Tese (Doutorado em Geografia Física) – Faculdade de Filosofia, Letras e Ciências Humanas (FFLCH), Universidade de São Paulo, São Paulo.

Keshava, N. (2003). A survey of spectral unmixing algorithms. *Lincoln Laboratory Journal, 14*, 55–78.

Keshava, N., & Mustard, J. F. (2002). Spectral unmixing. *IEEE Processing Magazine, 19*(2), 44–57.

Kruse, F., Lefkoff, A. B., Boardman, J. W., Heidebrecht, K. B., Shapiro, A. T., Barloon, P. J., & Goetz, A. F. (1993). The spectral image processing system (SIPS) – interactive visualization and analysis of imaging spectrometer data. *Remote Sensing of Environment, 44*, 145–163.

Lobell, D. B., & Asner, G. P. (2004). Cropland distributions from temporal unmixing of MODIS data. *Remote Sensing of Environment, 93*(3), 412–422.

Lu, D., Moran, E., & Batistella, M. (2003). Linear mixture model applied to Amazonian vegetation classification. *Remote Sensing of Environment, New York, 87*, 456–469.

Mcgwire, K., Minor, T., & Fenstermaker, L. (2000). Hyperspectral mixture modeling for quantifying sparse vegetation cover in arid environments. *Remote Sensing of Environment, 72*(3), 360–374.

Mertes, L., Smith, M., & Adams, J. (1993). Estimating suspended sediment concentrations in surface waters of the Amazon river wetlands from Landsat images. *Remote Sensing of Environment, 43*(3), 281–301.

Novo, E. M., & Shimabukuro, Y. E. (1997). Identification and mapping of the Amazon floodplain habitats using a mixing model. *International Journal of Remote Sensing, 18*(3), 663–670.

Peddle, D., Hall, F., & Ledrew, E. (1999). Spectral mixture analysis and geometric-optical reflectance modeling of boreal forest biophysical structure. *Remote Sensing of Environment, 67*(3), 288–297.

Poulet, F., & Erad, S. (2004). Nonlinear spectral mixing: Quantitative analysis of laboratory mineral mixtures. *Journal of Geophysical Research, 109*(E2), 1–12.

Quarmby, N., Townshend, J. R., Settle, J. J., & White, K. H. (1992). Linear mixture modelling applied to AVHRR data for crop area estimation. *International Journal of Remote Sensing, 13*(3), 415–425.

Quintano, C., Fernández-Manso, A., Shimabukuro, Y. E., & Pereira, G. (2012). Spectral unmixing. *International Journal of Remote Sensing, 33*, 5307–5340.

Quintano, C., Fernández-Manso, A., Fernández-Manso, O., & Shimabukuro, Y. E. (2006). Mapping burned areas in Mediterranean counties using spectral mixture analysis from a uni-temporal perspective. *International Journal of Remote Sensing, 27*(4), 645–662.

Quintano, C., Shimabukuro, Y. E., Fernández, A., & Delgado, J. A. (2005). A spectral unmixing approach for mapping burned areas in Mediterranean countries. *International Journal of Remote Sensing (Print), 26*(7), 1493–1498.

Richardson, A. J., & Wiegand, C. L. (1977). Distinguishing vegetation from soil background information. *Photogrammetric Engineering and Remote Sensing, New York, 44*, 1541–1552.

Roberts, D. A., Numata, I., Holmes, K., Chadwick, O., Batista, G., & Krug, T. (2002). Large area mapping of land-cover change in Rondônia using multitemporal spectral mixture analysis and decision tree classifiers. *Journal of Geophysical Research, 107*(D20), 40.001–40.017.

Rudorff, C., Novo, E., & Galvao, L. (2006). Spectral mixture analysis of inland tropical Amazon floodplain waters using EO-1 Hyperion. In *IEEE international symposium on geoscience and remote sensing*. (pp. 128–133). Denver, CO, USA: IEEE IGARSS.

Settle, J., & Campbell, N. (1998). On the errors of two estimators of subpixel fractional cover when mixing is linear. *IEEE Transactions on Geoscience and Remote Sensing, 36*(1), 163–170.

Settle, J., & Drake, N. (1993). Linear mixing and the estimation of ground cover proportions. *International Journal of Remote Sensing, 14*(6), 1159–1177.

Shimabukuro, Y. E., Holben, B. N., & Tucker, C. J. (1994). Fraction images derived from NOAA AVHRR data for studying the deforestation in the Brazilian Amazon. *International Journal of Remote Sensing, 15*(3), 517–520.

Shimabukuro, Y. E., Novo, E. M. L. M., & Ponzoni, F. J. (1998). Índice de vegetação e modelo de mistura espectral no monitoramento do Pantanal. *Pesquisa Agropecuária Brasileira (PAB)*, 33, 1729–1737.

Shimabukuro, Y. E., Almeida-Filho, R., Kuplich, T. M., & Freitas, R. M. (2008). Use of Landsat TM fraction images to quantify the optical and SAR data relationships for land cover discrimination in the Brazilian Amazonia. *International Journal of Geoinformatics, 4*(1), 57–63.

Van Der Meer, F. (1995). Spectral unmixing of Landsat Thematic Mapper data. *International Journal of Remote Sensing, 16*(16), 3189–3194.

Index

Y. E. Shimabukuro, F. J. Ponzoni, *Spectral Mixture for Remote Sensing*,
Springer Remote Sensing/Photogrammetry,
https://doi.org/10.1007/978-3-030-02017-0

Printed by Printforce, the Netherlands